Thomas Gozlinski

Local Probing of a Superconductor's Quasiparticles and Bosonic Excitations with a Scanning Tunnelling Microscope

Experimental Condensed Matter Physics
Band 32

Herausgeber
Physikalisches Institut
Prof. Dr. David Hunger
Prof. Dr. Alexey Ustinov
TT-Prof. Dr. Philip Willke
Prof. Dr. Wolfgang Wernsdorfer
Prof. Dr. Wulf Wulfhekel

Eine Übersicht aller bisher in dieser Schriftenreihe erschienenen Bände finden Sie am Ende des Buchs.

Local Probing of a Superconductor's Quasiparticles and Bosonic Excitations with a Scanning Tunnelling Microscope

by
Thomas Gozlinski

KIT Scientific Publishing

Karlsruher Institut für Technologie
Physikalisches Institut

Local Probing of a Superconductor's Quasiparticles and Bosonic Excitations with a Scanning Tunnelling Microscope

Zur Erlangung des akademischen Grades eines Doktors der Naturwissenschaften von der KIT-Fakultät für Physik des Karlsruher Instituts für Technologie (KIT) genehmigte Dissertation

von Thomas Gozlinski, M.Sc.

Tag der mündlichen Prüfung: 11. November 2022
Referent: Prof. Dr. Wulf Wulfhekel
Korreferent: Prof. Dr. Jörg Schmalian

Impressum

KIT Scientific Publishing

Karlsruher Institut für Technologie (KIT)
KIT Scientific Publishing
Straße am Forum 2
D-76131 Karlsruhe

KIT Scientific Publishing is a registered trademark
of Karlsruhe Institute of Technology.
Reprint using the book cover is not allowed.

www.ksp.kit.edu

Print on Demand 2023 – Gedruckt auf FSC-zertifiziertem Papier

ISSN 2191-9925
ISBN 978-3-7315-1279-0
DOI 10.5445/KSP/1000155716

Contents

1 Introduction 1

2 Superconductivity 5
2.1 Ginzburg-Landau Theory 5
2.1.1 The Ginzburg-Landau Parameter 7
2.1.2 Flux Pinning and Flux Flow 9
2.1.3 Anderson-Higgs Mechanism 10
2.2 Microscopic Theory of Conventional Superconductivity 12
2.2.1 Bardeen-Cooper-Schrieffer (BCS) Theory 12
2.2.2 Bogoliubov-de Gennes Theory 14
2.2.3 Migdal-Eliashberg Theory 15
2.3 Multiband Superconductivity 19
2.4 Mott Insulators 20
2.4.1 Granular Superconductors 22
2.4.2 Cuprates 23
2.5 Unconventional Superconductivity 23
2.5.1 Order Parameter Symmetry 23
2.5.2 Macroscopic and Microscopic Description of Magnetic Response in a Metal 26
2.5.3 Spin-Fermion Model 28
2.6 Quantum Tunnelling in NIS Junctions 31
2.6.1 Theoretical Framework of the Tunnelling Geometry 32
2.6.2 Elastic Tunnelling 33
2.6.3 Inelastic Tunnelling 35

3 Experimental Setup and Methods 39
3.1 Scanning Tunnelling Microscopy and Spectroscopy 39
3.2 Low Temperature Scanning Tunnelling Microscope 43
3.2.1 Dilution Scanning Tunnelling Microscope 44
3.2.2 Joule-Thomson Scanning Tunnelling Microscope 47
3.3 Tip and Sample Preparation 47
3.3.1 Tip Preparation 47

3.3.2 Sputtering and Annealing 49
3.3.3 In-situ Cleaving 50

4 Bosonic Excitation Spectrum in $Bi_2Sr_2CaCu_2O_{8+\delta}$ and $YBa_2Cu_3O_{6+x}$ . . 51
4.1 Structural and Electronic Properties 51
4.1.1 Structural Properties of Y123 and Bi2212 51
4.1.2 Unconventional Pairing in Y123 and Bi2212 53
4.2 Inelastic Tunnelling in Spin-Fluctuation Driven Superconductors . . 56
4.3 Separation of Elastic and Inelastic Tunnelling Events 58
4.4 Determining the Dispersion of the Bosonic Glue 64
4.5 Summary 67

5 Yu-Shiba-Rusinov States and Long-lived Bosonic Excitations in Granular Aluminium 69
5.1 Relation between T_c and Oxygen Partial Pressure 70
5.2 Microscopic Grain Coupling and the Superconducting Gap 72
5.2.1 Pure Al Film 73
5.2.2 Oxygen Poor GrAl Film 73
5.2.3 Oxygen Rich GrAl Film 77
5.3 Spectroscopic Signs of YSR states 79
5.4 Spectroscopic Signs of Long-lived Plasma or Higgs modes 81
5.5 *LC* Resonator Model for Plasmon Modes 87
5.6 Tien-Gordon Model for Spontaneous Plasmon Emission 90
5.7 First Results on Oxidized Polycrystalline Al 92
5.8 Summary 94

6 Multiflux Vortex States in Bulk Pb 97
6.1 Meißner and Intermediate State 97
6.2 Emergence of Isolated Vortices 100
6.2.1 Normal Single-Flux Vortices 100
6.2.2 Anomalous Vortices 105
6.3 Tunnelling into Caroli-de Gennes-Matricon States 108
6.3.1 Determination of Vortex Winding Numbers 109
6.4 Evidence of Multiflux Vortices 110
6.5 Vortex Interactions beyond the Ginzburg-Landau Limit 112
6.6 Summary 115

7 Quasiparticle and Flux Trapping in Topological Defects of a Two-band Superconductor 117
7.1 Stacking Fault Tetrahedra 118
7.1.1 Stacking Fault Tetrahedra in Pb(111) 120

7.2 Band Filter Effect and Quantum Well States 122
7.3 Electronic Edge State 127
7.4 Summary 132

8 Conclusion and Outlook 133

Bibliography 137

Acronyms 161

Notation 165

List of Figures 167

List of Tables 171

Appendix A Deconvolution using the Gold Algorithm 175

Appendix B Supplementary GrAl Data 177

Appendix C 3D Eilenbenberger Equations 179
C.1 Intermediate Results 183

Appendix D Supplementary Pb Vortex Data 185
D.1 Bias and Magnetic Field Dependence of CdGM LDOS 185
D.2 Self-consistent Solution of the Gap Equation 186

Appendix E Supplementary Pb Defect Data 187

Acknowledgements 189

1 Introduction

Superconductors are materials with two very unique characteristics: They are able to transmit a current without electrical resistance and can completely expel external magnetic fields from their interior (Meißner-Ochsenfeld effect). Unfortunately, all superconducting materials we know so far, only show these properties under extreme conditions, i.e. very low temperature or large pressure, which currently limits their application to large scale science projects or complex medical imaging techniques. Their potential is, however, so great and diverse that superconductivity at ambient conditions is a long-sought holy grail of condensed matter physics. Key technologies that would follow such a discovery range from lossless electrical power transmission and storage to less consumptive transportation alternatives or powerful quantum computation [1].

The history of fundamental research in superconductivity is a prime example of how scientific progress often happens in quantum leaps and in areas where we least expect it. When we think of good conductors, we think of silver, gold or platinum. The first discovered superconductor was mercury. In 1911, Kammerlingh Onnes measured a vanishing electrical resistance below a critical temperature of 4.2 K [2–4]. While mercury is a metal at room-temperature, it is a rather bad metal when compared with gold, silver or platinum. Things became only more absurd with the discovery of the first high temperature superconductor (HTS). Following the discovery of superconductivity in mercury, scientists found superconductivity in more and more metals and metallic alloys or compounds. Bernd Matthias was convinced that he had found a pattern after which materials did or did not show superconductivity which became known as Matthias' rules [5]. Loosely speaking, they state that superconductivity favours transition metals over simple metals, a high electronic density of states (DOS), high symmetry and that it does not like oxygen, magnetism or insulating phases [6]. The cuprate HTS discovered by Bednorz and Müller in 1986 [7], an antiferromagnetic insulator with copper-oxide planes, did not break one or two, but, at least to some extent, every single one of these rules. Clearly, these types of superconductors with critical temperatures of up to 138 K [8] were fundamentally different and to this day there is no complete microscopic theory that can explain the unconventional superconductivity in these systems. But the history of theory of superconductivity is also a story of great success and the currently world-record setting high-pressurized hydrides [9, 10] are a testimony to our good understanding of conventional superconductors [11]. But why is it that "bad metals" make good superconductors? In order to understand this, we have to go back to the early days of superconductivity theory.

The 1950s, roughly 40 years after superconductivity was found in mercury, turned out to be some of the most important years for the history of superconductivity, that still influence the way we understand this phenomenon today. The isotope effect [12], i.e. the dependence of the critical temperature T_c and critical field H_c on the atomic mass in isotopes of the same element, led Fröhlich to believe that electron-lattice interactions are of utmost importance for the superconducting state to form [13]. He could show that although the static Coulomb interaction between two electrons is repulsive, a net attractive force between two electrons can be mediated by the dynamic interaction of electrons with the crystal lattice [13, 14]. A few years later, Cooper could show that the Fermi sea of electrons is unstable against even the smallest attractive interaction, which leads electrons to form pairs, the so-called Cooper pairs [15]. A year after, in 1957, the Nobel Prize winning work of Bardeen, Cooper and Schrieffer [16] finally tied all the threads together to a coherent microscopic theory of conventional superconductivity, that could describe the relation between microscopic parameters like the binding energy of a Cooper pair and macroscopic quantities like the critical temperature. With it, the vanishing resistance, the Meißner effect and the jump in the specific heat at T_c could all be explained from fundamental interactions of electrons in solids. With the extension to a strong-coupling theory [17], we also learned that there is in principle no natural upper limit for T_c [18] and that a stronger coupling between electrons and lattice vibrations does not only lead to an increased normal state conductivity due to electron scattering but can also lead to higher critical temperatures.

It was during this time period, that the Norwegian physicist Giaever developed the technique of electron tunnelling spectroscopy [19], which later earned him a Nobel Prize. From there on, tunnelling experiments, tunnelling theory and the theory of superconductivity developed side by side. Today, electron tunnelling spectroscopy can be performed locally by a scanning tunnelling microscope (STM) and state-of-the-art electronics and cryogenics enable a deep look into the electronic structure of superconductors with unprecedented energy resolution. A detailed introduction into the theory of superconductivity and electron tunnelling is given in Chap. 2, followed by a description of the experimental setup and methods in Chap. 3.

Based on recent advances in our understanding of superconductivity and related phenomena, we want to go back in time and use the capability of a state-of-the-art STM to revisit superconductors that still leave room for undiscovered things. We start with two popular representatives of the cuprate family of HTSs (Chap. 4). With granular aluminium (grAl) (Chap. 5), we then turn to a granular superconductor with a T_c enhancement mechanism that is not yet understood, but that was already discovered in 1966 [20]. Lastly, we want to have a look at the conventional superconductor lead (Pb) (Chap. 6 and 7).

Antiferromagnetic spin fluctuations are one of the hottest contenders for the exchange boson in unconventional superconductors. With the recent theoretical progress in the spin-fermion model [21, 22] and tunnelling experiments revealing bosonic spectra [23–25] that show remarkable resemblence to the spin resonance seen in inelastic neutron scattering (INS) [26], we slowly reach a consensus on the pairing mechanism in cuprate superconductors. J. Jandke and P. Hlobil could show in a combined experimental and theoretical study [27, 28] that the spectral density of the pairing boson is also accessible through the inelastic part of a tunnelling current. STM provides conditions under which an effective Eliashberg function, which is proportional to the spectral density of the pairing boson, can be extracted. The extraction of the effective Eliashberg function was in that case done for an iron-based superconductor with a full gap. In this work, the aim is to apply the formalism to two cuprate superconductors with a d-wave symmetry of the superconducting gap.

GrAl recently reemerged in the superconducting qubit community as a suitable superinductor in Al/AlO_x based superconducting circuits [29–31] and as a template for magnetic-field resilient qubit fabrication [32,33]. Not only are excess quasiparticles identified as a limiting loss mechanism in these circuits, transport measurements in grAl near its metal-to-insulator transition (MIT) also hint at localized magnetic moments that might be responsible for the $1/f$ flux noise in Al/AlO_x based junctions [34,35]. For some reason, a thorough STM study of this material is (to our knowledge) missing to this date. With our local probe of the quasiparticles on a microscopic scale, we are sensitive to spin scattering, which should manifest as a pair of Yu-Shiba-Rusinov (YSR) states below T_c. The aim in the scope of this thesis is to provide a detailed STM study of the material with special focus on the possible origins of microwave losses.

Multiband superconductivity is still a relatively young branch of superconductivity. Magnesium diboride (MgB_2) with a T_c of 39 K [36] is the first superconductor for which two-band superconductivity was confirmed in 2001/2002 [37, 38]. Two-band superconductivity was also recently demonstrated for lead (Pb) [39]. In MgB_2, the existence of two superconducting condensates has interesting implications for the intermediate state and the dynamics of vortices that has been labeled type-1.5 superconductivity [40]. We are interested if a similar thing happens in Pb. Bulk Pb is not expected to host vortices in its intermediate state at temperatures close to its T_c of 7.2 K as it is categorized as a traditional type-I superconductor [41]. However, Pb is the elemental superconductor that is closest to the Bogomol'nyi point that separates type-I and type-II, which is demonstrated in several experiments that initiate a type-I to type-II transition through a reduction of the effective coherence length [42, 43] or an increase in the magnetic penetration depth [44, 45]. Additionally, with our base measurement temperature of $25\,\mathrm{mK}$, we are far below T_c and in relatively unexplored territory. This is more of an explorative study that is solely motivated by scientific curiosity. We thus find it more instructing to lead the reader through the thought process behind consecutive measurements along the way and motivate them by the newly gained insights at each waypoint.

2 Superconductivity

This chapter summarizes fundamental theoretical concepts of conventional and unconventional superconductivity and gives an introduction into tunnelling theory. The first section features the Ginzburg-Landau theory of superconductivity. It is followed by a section that covers important results of the microscopic theory of conventional superconductors. The sections on multiband superconductivity and Mott insulators familiarize the reader with superconductivity under unusual circumstances. The fourth section is dedicated to unconventional superconductivity with special focus on the cuprates and a spin-fluctuation mediated pairing mechanism. In the last section, tunnelling theory, which is essential for the spectroscopic measurement of superconductors, is introduced.

2.1 Ginzburg-Landau Theory

A powerful macroscopic theory to describe superconductors at temperatures $T \sim T_c$, i.e. close to the transition temperature, is the *Ginzburg-Landau (GL) theory*. It is a generalization of the *Landau theory* of second-order phase transitions, that enables one to study the low-temperature ordered state in the presence of spatial inhomegeneities. In the specific case of a superconductor, this inhomogeneity can be induced by electric currents or magnetic fields. The GL theory is a scalar field theory that describes the second-order phase transition through a thermodynamic parameter $\psi(\boldsymbol{r})$, which is called the order parameter. ψ continuously changes from zero to a finite value upon deceeding the transition (or critical) temperature T_c, like in Fig. 2.1(a). The inhomegeneity of the system is treated through the inclusion of gradient terms of this scalar field. The lowest order of scalar field theory, in which these gradient terms play a role, is a ϕ^4-theory. For $T \approx T_c$, the free-energy of the system is expanded as a power series of the order parameter up to terms of fourth power in ψ. The free energy functional reads

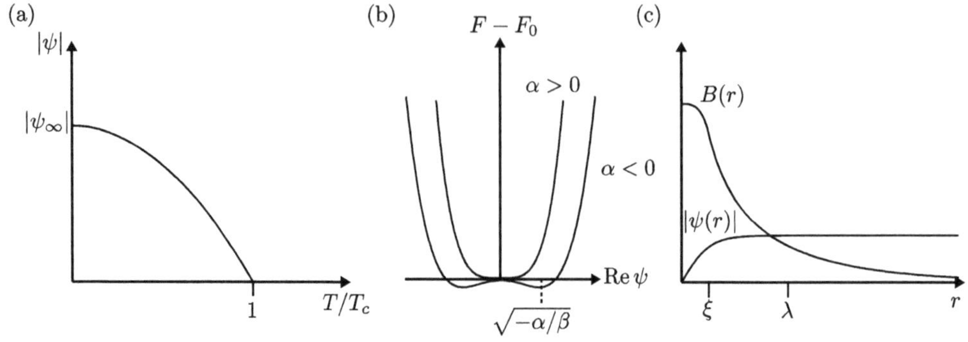

Figure 2.1: Ground State in GL Theory: (a) Temperature dependence of the order parameter. (b) Free energy landscape in the normal $\alpha > 0$ and spontaneously symmetry broken state $\alpha < 0$. (c) Magnetic field and order parameter profile around a vortex in a type-II superconductor. The magnetic field decays exponentially on the lengthscale λ while the order parameter recovers on the lengthscale ξ.

$$F[\psi(\boldsymbol{r}), \boldsymbol{A}(\boldsymbol{r})] = F_0 + \int \mathrm{d}r^3 \left[\alpha|\psi(\boldsymbol{r})|^2 + \frac{\beta}{2}|\psi(\boldsymbol{r})|^4 + \right.$$
$$\left. + \frac{1}{2m^*}|(-i\hbar\nabla - q\boldsymbol{A}(\boldsymbol{r}))\psi(\boldsymbol{r})|^2 + \frac{|\boldsymbol{B}(\boldsymbol{r})|^2}{2\mu_0} \right]. \tag{2.1}$$

F_0 is the free energy of the normal state above T_c, $\boldsymbol{B} = \nabla \times \boldsymbol{A}$ is the magnetic field and $\boldsymbol{A}$ the corresponding magnetic vector potential. For a superconductor, the charge and mass of each charge carrier is $q = 2e$ and $m^* = 2m_e$, because the supercurrent is carried by Cooper pairs, that are made up of two electrons. The free energy landscape is shown in Fig. 2.1(b). For $\alpha > 0$, the free energy functional is minimized by a single-valued ψ. For $\alpha < 0$, a global $U(1)$ symmetry is spontaneously broken and the energy landscape has more than one global minimum. In fact, if one extends the scalar field to the complex plane, i.e. $\psi = |\psi|e^{i\varphi}$, the energy landscape forms a *mexican hat potential*. The sign change of α at T_c is ensured by writing $\alpha(t) = \alpha_0(t-1)$ with $t = T/T_c$. The minimization of Eq. (2.1) with respect to the order parameter ψ and the vector potential $\boldsymbol{A}$ yields the two *GL equations*:

$$\frac{1}{2m^*}(-i\hbar\nabla - q\boldsymbol{A})^2\psi + \alpha\psi + \beta|\psi|^2\psi = 0\,, \tag{2.2}$$
$$\boldsymbol{j}_s = \frac{1}{\mu_0}\nabla \times \boldsymbol{B} = i\frac{q\hbar}{2m^*}\left([\nabla\psi^*]\,\psi - \psi^*\nabla\psi\right) - \frac{q^2}{m^*}|\psi|^2\boldsymbol{A}. \tag{2.3}$$

The dependence of $\boldsymbol{r}$ was dropped for a shorter notation. In the field free, homogeneous case ($\boldsymbol{A} = 0$, $\nabla\psi = 0$), the first GL equation yields the ground state

$$|\psi|^2 = |\psi_\infty|^2 = -\frac{\alpha(t)}{\beta}. \tag{2.4}$$

Allowing for spatial variations of the order parameter $\psi = \psi_\infty + \delta\psi(\boldsymbol{r})$, one finds from Eq. (2.2) that the deviations from the homegeneous ground state decay on a characteristic length scale

$$\xi(t) = \sqrt{\frac{\hbar^2}{2m^*|\alpha(t)|}} \propto \sqrt{\frac{1}{1-t}}, \quad t < 1, \tag{2.5}$$

which is the GL coherence length. A second characteristic length scale is obtained by taking the curl of the second GL equation (2.3), which states that magnetic fields that penetrate the superconductor are screened by induced currents and are consequently exponentially damped on the length scale

$$\lambda(t) = \sqrt{\frac{m^*}{\mu_0|\psi(t)|^2 q^2}} \propto \sqrt{\frac{1}{1-t}}, \quad t < 1, \tag{2.6}$$

which is called the London penetration depth. Another important implication of Eq. (2.3) is that the magnetic flux, that penetrates the superconductor, is quantized. One can obtain this result by integrating the second GL equation along a closed line far away from the region where the field lines penetrate, so that the order parameter varies only in its phase $\nabla\psi = i\psi\nabla\varphi(\boldsymbol{r})$ and that the net current through this loop is zero: $\oint \mathrm{d}\boldsymbol{s}\boldsymbol{j}_s = 0$. The single-valuedness of the order parameter then demands

$$\oint \mathrm{d}\boldsymbol{s}\nabla\varphi = \frac{q}{\hbar}\Phi = 2\pi n. \tag{2.7}$$

The magnetic flux $\Phi = \oint \mathrm{d}\boldsymbol{s}\boldsymbol{A}$ is hence quantized in units of the magnetic flux quantum $\Phi_0 = h/2e$.

2.1.1 The Ginzburg-Landau Parameter

It turns out out, that superconductors can be classified into two types based on their response to a magnetic field, which is determined through the ratio of the two characteristic length scales

$$\kappa(t) \equiv \frac{\lambda(t)}{\xi(t)}. \tag{2.8}$$

κ is called the GL parameter. A superconductor with $\kappa < 1/\sqrt{2}$ is classified as type-I and a superconductor with $\kappa > 1/\sqrt{2}$ as type-II. Through the Meißner-Ochsenfeld effect, a superconductor tries to actively expel an externally applied field by its screening currents. At a certain field strength H_c, however, the associated energy that is needed to realize a superconducting ground state with $\boldsymbol{B}_{\mathrm{int}} = 0$ is simply larger than the free energy of the normal state and the system returns to the normal state

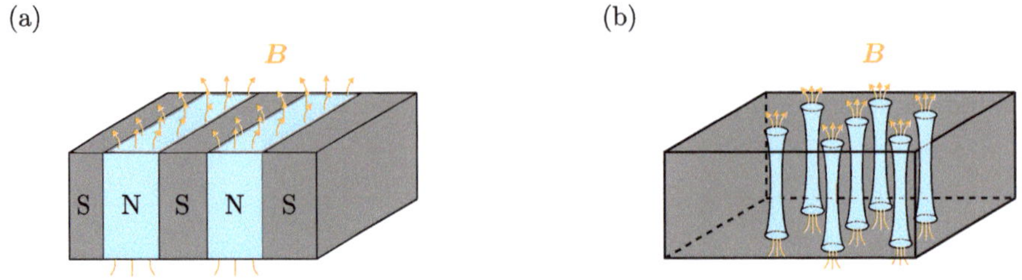

Figure 2.2: Intermediate and Mixed Phase: (a) Schematic domains of a type-I superconductor in the intermediate state. (b) Schematic Abrikosov lattice of a type-II superconductor in the mixed phase.

via a first-order phase transition. For a type-I superconductor of finite dimension the phase transition is known to occur locally, which leads to an intermediate phase, in which large normal and superconducting regions coexist. This scenario is depicted in Fig. 2.2(a). For a type-II superconductor, there exists an additional Shubnikov phase at $H_{c,1} < H < H_{c,2}$, in which the superconductor favors flux entry in the form of homogeneously distributed flux tubes carrying a single flux quantum Φ_0, each. These flux tubes are known as vortices, and the lattice they form in a type-II superconductor is called the Abrikosov lattice, which is shown in Fig. 2.2(b). Below the lower critical field $H_{c,1}$ the superconductor is again in the Meißner phase and above the upper critical field $H_{c,2}$ it is in the normal phase. To illustrate the meaning of the coherence length and London penetration depth, the radial field and order parameter profile of a vortex in a type-II superconductor is schematically shown in Fig. 2.1(c).

One can understand the formation of these two transitional phases of a superconductor by taking a look at the energy cost of a domain wall. For a strong type-I superconductor, $\xi \gg \lambda$. Therefore, a single domain wall can lead to a large region in which the energy cost for expelling the magnetic field is already large but the density of the superconducting condensate $|\psi|^2$ is only slowly approaching its bulk value $|\psi_\infty|^2$. The associated energy for hosting one domain wall is positive, so the system tries to minimize the surface area of domain walls. In contrast, a strong type-II superconductor with $\xi \ll \lambda$ is less effective in screening the penetrating magnetic field. But there, the energy gain due to the superconducting condensation outweighs the energy cost from screening. As a result, the associated energy for a single domain wall is negative and the system tries to maximize the surface area of domain walls. This is realized by a close-packed array of single-flux vortices.

2.1.2 Flux Pinning and Flux Flow

The flux lines inside a superconductor are a constant source of dissipation. In the presence of an electric current density $\boldsymbol{j}$ and magnetic field $\boldsymbol{B}$, the Lorentz force density acting on the flux lines is given by [46, p.163]

$$\boldsymbol{f} = \boldsymbol{j} \times \boldsymbol{B}. \tag{2.9}$$

The movement of vortices or larger normal conducting regions through the crystal lattice equals a normal conducting current in their core, which is responsible for the dissipation. In the *flux flow* regime, the stopping force of the vortices can be described by a viscous drag coefficient [46, p.166]

$$\eta = \frac{j\Phi}{v_L}, \tag{2.10}$$

where v_L is the velocity of the vortex with respect to the lattice and Φ is the flux threaded through it.

In the absence of a transport current, the magnetic field lines enter a finite superconductor from the sides and are propelled to the inside by the magnetic field gradient. Due to the movement of their normal conducting cores, their motion is damped by dissipating currents and eventually stops. An array of vortices may also be moved by an electric field in their normal core induced by magnetic induction according to $\nabla \times \boldsymbol{E} = -\partial \boldsymbol{B}/\partial t$ [47]. Upon stopping the magnetic field ramp, the vortex motion is again damped by the viscous behaviour according to Eq. (2.10), i.e. the dynamics do not stop instantaneously.

The case $v_L = 0$ for $j \neq 0$ is only reached for $\eta \to \infty$. This is a mechanism called *flux pinning*. So far, only a perfect, defect free superconductor was discussed. Inhomogeneities like point defects, inclusions, twin boundaries, stacking faults or dislocations are only a few crystal defects that can act as pinning centres. In general, for effective pinning, the defect should have the dimension of $\sim \xi$ or $\sim \lambda$ [46, p.163]. Point defects are therefore only effective in a few high temperature superconductors, in which ξ is of the same order as the lattice constant. Extended defects like dislocations, however, can also locally suppress the order parameter in elemental superconductors. From an energetic consideration, it is favorable for the flux to thread through the defect than through the superconducting bulk next to the defect, where the order parameter is already fully recovered. This local minimum of $F[\boldsymbol{A}]$ creates a pinning force on flux lines near it. In order to unpin the flux lines, the pinning force needs to be overcome by an overcompensating Lorentz force.

2.1.3 Anderson-Higgs Mechanism

The famous *Goldstone theorem*, which was born in the context of superconductivity by the works of Nambu [48], states that any spontaneous breaking of a continuous symmetry necessitates the emergence of massless (gapless) bosons, one for each generator of a broken symmetry [49, 50]. These bosons are referred to as *Nambu-Goldstone (NG) bosons*. Famous examples of NG modes in condensed matter physics are the magnons in ferromagnetic systems as a consequence of a broken rotational invariance in spin-space and acoustic phonons in crystals as a consequence of a broken rotational and translational invariance. The most original example, however, is the NG mode in a superconductor, which appears as a consequence of a broken global $U(1)$ symmetry.

A BCS superconductor is often casually said to spontaneously break the $U(1)$ gauge symmetry when entering the superconducting state. This is, however, not entirely correct because a gauge symmetry is not a real, physical symmetry, but merely a choice of parametrisation of the system[1][52]. Therefore, it cannot be spontaneously broken [53]. What is meant, is that the global part of a $U(1)$ phase rotation symmetry

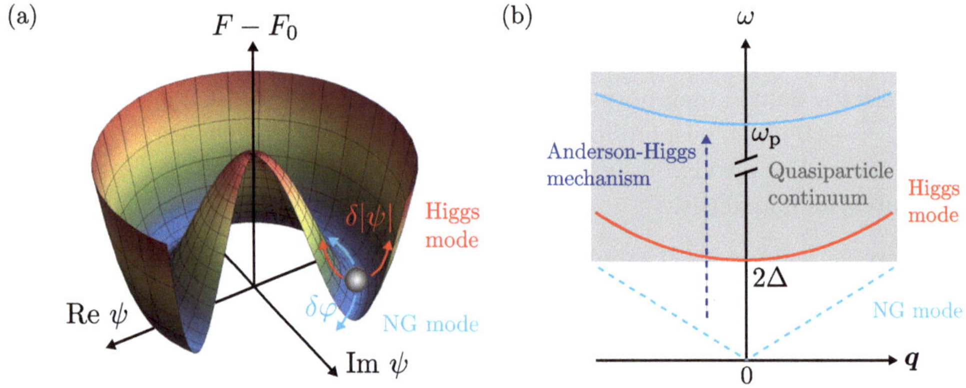

Figure 2.3: Anderson-Higgs Mechanism: (a) Spontaneous symmetry breaking of $U(1)$ leads to an extra degree of freedom, that is a massless NG mode (phase fluctuations), and a massive Higgs mode (amplitude fluctuations) of the order parameter ψ. (b) Dispersion $\omega(\boldsymbol{q})$ of the NG mode and the Higgs mode. The NG mode is "eaten up" in the gauge of the vector potential leading to massive photons that are gapped by the plasma frequency ω_p. The Higgs mode can only decay into the quasiparticle continuum for $|\boldsymbol{q}| > 0$. Both adapted from [51].

[1] The difference between a physical symmetry and a gauge symmetry can be understood as the difference between the rotation of an object and the rotation of the coordinate axes spanning the reference frame. While the first has a physical consequence, the second is only a description of the system in terms of different variables.

(which formally looks very much like a global gauge transformation) of the fermionic field operators is spontaneously broken. This symmetry is, however, a real, physical symmetry and tied to the number of fermions (or volume of the Fermi surface) in the system, which is not conserved anymore when electrons start to pair [54, 55]. As a consequence of the spontaneously broken symmetry, the system acquires the superconducting gap $|\Delta|$ as an order parameter and an infinite number of degenerate ground states related by a phase φ, i.e. $\Delta = \hat{\Delta}e^{i\varphi}$, that is necessarily universal throughout the system in its ground state. That this phase has physical implications and is not the freedom of a gauge is reflected in the DC Josephson effect. A superconductor is a special case, in which the matter field and gauge field, that is the electromagnetic vector potential $\boldsymbol{A}$, are coupled. Meissner effect, flux quantization, Josephson effects are all phenomena based on this principle [56].

The $U(1)$ symmetry describes the macroscopic phase of the order parameter, which has a preferred value in the ordered state. This is shown in the mexican-hat potential in Fig. 2.3(a). It turns out, that if such a symmetry can be associated with a gauge symmetry, like in the case of the electroweak interaction, then the massless NG mode is not visible, because it is "eaten up" by the gauge field and leads to massive bosons [57–61]. This is the so-called *Anderson-Higgs mechanism*. In the context of the electroweak interaction, these massive bosons are the W and Z bosons. In the context of superconductivity, the photons inside the superconductor gain a mass, which is ultimately responsible for the Meißner effect.

To see how this works, one can write down the GL functional for $T < T_c$ and insert the general ansatz $\psi = (\psi_\infty + \delta\psi(\boldsymbol{r}))e^{i\varphi(\boldsymbol{r})}$, that allows for phase and amplitude fluctuations of the order parameter. After some algebraic transformations one finds [62]

$$F[\delta\psi, \boldsymbol{A}] = F_0 + \int d^3r \left[-2\alpha\delta\psi^2 + \frac{1}{2m^*}\left|-i\hbar\nabla\delta\psi\right|^2 \right. \\ \left. -\frac{\alpha}{\beta}\frac{\hbar^2}{2m^*}\left|\nabla\varphi - \frac{q}{\hbar}\boldsymbol{A}\right|^2 + \frac{1}{2\mu_0}\left|\nabla\times\boldsymbol{A}\right|^2 \right]. \tag{2.11}$$

This functional is invariant under the gauge transformation

$$\boldsymbol{A}' := \boldsymbol{A} - \frac{\hbar}{q}\nabla\varphi, \quad \psi' := \psi e^{-i\varphi}. \tag{2.12}$$

By analyzing the form of Eq. (2.11), one recognizes two constant mass terms: The mass term for the amplitude fluctuations is the first term and the mass is -2α. The mass term for the field $\boldsymbol{A}$ is the third term and the mass is $-\alpha q^2/(2\beta m^*)$. As the microscopic theory shows, the mass term of the vector potential is equal to the plasma frequency $\omega_p = \sqrt{n_s q^2/(\varepsilon_0 m^*)}$ [57] and the amplitude fluctuations are gapped by 2Δ [63], which is the characteristic energy gap of fermions in a superconductor and real part of the order parameter. The dispersion relation of these two modes is also shown in Fig. 2.3(b). It becomes obvious, that the quasiparticle gap 2Δ becomes completely empty through the Anderson-Higgs mechanism and the consequent elevation of the NG mode. The amplitude mode, which is commonly referred to as the *Higgs mode* can only decay for $\boldsymbol{q} \neq 0$, because the energy transfer to quasiparticles must be at least 2Δ. The phase space for the decay of the Higgs mode into quasiparticles is, however, too large to observe the Higgs mode in most superconducting systems. According to a few theoretical studies, it should be possible to shrink the number of decay channels by decreasing the gap of the Higgs mode below 2Δ in superconductors with a charge-density wave [63] or disordered superconductors close to a superconductor-to-insulator transition (SIT) [64]. The phase mode is lifted in energy to the plasma frequency ω_p, which is far above the gap, at least in 3D bulk systems. In superconductors an excitation of the phase mode is, therefore, equivalent to the excitation of a plasmon.

2.2 Microscopic Theory of Conventional Superconductivity

This section gives a brief overview of important results of the Bardeen-Cooper-Schrieffer (BCS) theory for conventional superconductors [16] and the Migdal-Eliashberg theory for strong coupling superconductors [17, 65, 66].

2.2.1 BCS Theory

Given some attractive electron-electron interaction, the BCS formalism describes the pairing mechanism that leads to a superconducting condensate, the BCS ground state. It starts from a Fermi liquid theory of the normal state, i.e. the direct Coulomb interaction between electrons can be absorbed into single-particle excitations ("dressed" electrons) of the Fermi sea with energy $\epsilon_{\boldsymbol{k}}$ with respect to the chemical potential μ. The scattering process with the largest phase space, that destabilizes the Fermi liquid, is the one between a pair of electrons in the initial state $(\boldsymbol{k}' \uparrow, -\boldsymbol{k}' \downarrow)$ into the unprimed final state $(\boldsymbol{k} \uparrow, -\boldsymbol{k} \downarrow)$. The four-particle scattering vertex is described by the potential $V_{\boldsymbol{k},\boldsymbol{k}'}$ (cf. Fig. 2.4) and the reduced Hamiltonian in second quantization reads

$$H = \sum_{\boldsymbol{k}\sigma} \epsilon_{\boldsymbol{k}} c^\dagger_{\boldsymbol{k}\sigma} c_{\boldsymbol{k}\sigma} + \frac{1}{N} \sum_{\boldsymbol{k}\boldsymbol{k}'} V_{\boldsymbol{k}\boldsymbol{k}'} c^\dagger_{\boldsymbol{k}\uparrow} c^\dagger_{-\boldsymbol{k}\downarrow} c_{-\boldsymbol{k}'\downarrow} c_{\boldsymbol{k}'\uparrow}. \tag{2.13}$$

Here and in the rest of this subsection, we closely follow and adopt the equations from Ref. [62]. Bardeen, Cooper and Schrieffer realized that it is useful to reformulate a mean-field approximation of this Hamiltonian in terms of expectation values for pair-annihilation and pair-creation. This leads to the complex valued BCS gap parameter $\Delta_{\boldsymbol{k}}$ defined by the BCS equation

$$\Delta_{\boldsymbol{k}} = -\frac{1}{N}\sum_{\boldsymbol{k}'} V_{\boldsymbol{k}\boldsymbol{k}'}\langle c_{-\boldsymbol{k}'\downarrow}c_{\boldsymbol{k}'\uparrow}\rangle = -\frac{1}{N}\sum_{\boldsymbol{k}'} V_{\boldsymbol{k}\boldsymbol{k}'}\frac{\Delta_{\boldsymbol{k}'}}{2E_{\boldsymbol{k}'}}\tanh\frac{E_{\boldsymbol{k}'}}{2k_B T} \tag{2.14}$$

with the gapped quasiparticle (QP) spectrum

$$E_{\boldsymbol{k}} = \sqrt{\epsilon_{\boldsymbol{k}}^2 + \Delta_{\boldsymbol{k}}^2}. \tag{2.15}$$

$\Delta_{\boldsymbol{k}}$ describes the mean binding energy of a pair of electrons with opposite momenta and spin. Its complex conjugate describes the mean binding energy of a pair of holes with opposite momenta and spin. Bogoliubov [67] and Valatin [68] could show independently that the mean field Hamiltonian acquired by Bardeen, Cooper and Schrieffer can be diagonalized by a linear transformation that mixes electron and hole excitations. It then simplifies to

$$H = E_{\mathrm{BCS}} + \sum_{\boldsymbol{k}\sigma} E_{\boldsymbol{k}}\gamma_{\boldsymbol{k}\sigma}^\dagger\gamma_{\boldsymbol{k}\sigma} \tag{2.16}$$

with the new fermionic creation and annihilation operators for the Bogoliubov particles (or Bogoliubons)

$$\begin{pmatrix}\gamma_{\boldsymbol{k}\uparrow}\\ \gamma_{-\boldsymbol{k}\downarrow}^\dagger\end{pmatrix} = \begin{pmatrix}u_{\boldsymbol{k}}^* & -v_{\boldsymbol{k}}\\ v_{\boldsymbol{k}}^* & u_{\boldsymbol{k}}\end{pmatrix}\begin{pmatrix}c_{\boldsymbol{k}\uparrow}\\ c_{-\boldsymbol{k}\downarrow}^\dagger\end{pmatrix} \tag{2.17}$$

and the superconducting ground state energy

$$E_{\mathrm{BCS}} = \sum_{\boldsymbol{k}}(\epsilon_{\boldsymbol{k}} - E_{\boldsymbol{k}} + \Delta_{\boldsymbol{k}}\langle c_{\boldsymbol{k}\uparrow}^\dagger c_{-\boldsymbol{k}\downarrow}^\dagger\rangle). \tag{2.18}$$

The BCS ground state is thus a vacuum of Bogoliubons:

$$\gamma_{\boldsymbol{k}\sigma}\left|\psi_{\mathrm{BCS}}\right\rangle = 0. \tag{2.19}$$

From Eq. (2.16) it becomes obvious why Eq. (2.15) was called the QP spectrum with the characteristic gap $\Delta_{\boldsymbol{k}}$. The superconducting gap is most notably seen in the DOS of the Bogoliubov QPs. We follow again Bardeen, Cooper and Schrieffer and assume a constant attractive pair interaction for electrons and holes in a thin shell around the Fermi energy of width $2\omega_D$:

$$V_{\boldsymbol{k}\boldsymbol{k}'} = \begin{cases} -V_0\,, & |\epsilon_{\boldsymbol{k}}|\,,|\epsilon_{\boldsymbol{k}'}| < \omega_D \\ 0\,, & \text{otherwise.}\end{cases} \tag{2.20}$$

ω_D, the Debye frequency, is the natural cut-off frequency for an attraction mechanism mediated by phonons. Under this assumption, $\Delta_{\boldsymbol{k}} \to \Delta$ and the DOS of Bogoliubov QPs is given by

$$\nu_{\mathrm{BCS}}(\epsilon) = \begin{cases} \nu_{\mathrm{n}}^F \frac{|\epsilon|}{\sqrt{\epsilon^2 - \Delta^2}} & \text{for } |\epsilon| > \Delta\,, \\ 0, & \text{for } |\epsilon| < \Delta. \end{cases} \tag{2.21}$$

Here, ν_{n}^F is the DOS at the Fermi energy in the normal state. In the BCS approximation (Eq. (2.20)), the gap equation (2.14) can be solved analytically and yields for $T = 0$

$$\Delta(0) \approx 2\omega_D \mathrm{e}^{-\frac{1}{\nu_{\mathrm{n}}^F V_0}}. \tag{2.22}$$

It is related to the critical temperature T_c via the approximate relation

$$\frac{\Delta(0)}{k_B T_c} = 1.764. \tag{2.23}$$

Gorkov could derive the Ginzburg-Landau equations from the BCS equations and identified the position and temperature dependent gap $\Delta(\boldsymbol{r}, T)$ as the order parameter $\psi(\boldsymbol{r}, T)$ of a BCS superconductor [69] with the same $(1 - T/T_c)^{1/2}$ dependence close to T_c.

2.2.2 Bogoliubov-de Gennes Theory

Often, one is interested in local inhomogeneities of the superconducting state. The Bogoliubov-de Gennes (BdG) formalism provides a microscopic description that enables us to study defects, like e.g. vortices. One starts from the BCS mean-field Hamiltonian and makes it spatially dependent. The QPs are described by the two-component spinor [62]

$$|\Psi_{\boldsymbol{k}}\rangle = \begin{pmatrix} c_{\boldsymbol{k}\uparrow}^\dagger \\ c_{-\boldsymbol{k}\downarrow} \end{pmatrix} |\psi_{\mathrm{BCS}}\rangle \tag{2.24}$$

and by the Hamiltonian

$$H_{\mathrm{BdG}}(\boldsymbol{k}) = \begin{pmatrix} \epsilon_{\boldsymbol{k}} & -\Delta_{\boldsymbol{k}} \\ -\Delta_{\boldsymbol{k}}^* & -\epsilon_{\boldsymbol{k}} \end{pmatrix} \tag{2.25}$$

with the eigenvalues $\pm E_{\boldsymbol{k}}$ from Eq. (2.15). The spatial dependence is obtained through a Fourier transformation of the single QP excitations with momentum $\boldsymbol{k}$ [62]:

$$\Psi(\boldsymbol{r}) = \frac{1}{\sqrt{N}} \sum_{\boldsymbol{k}} \mathrm{e}^{i\boldsymbol{k}\boldsymbol{r}} \left|\Psi_{\boldsymbol{k}}\right\rangle , \tag{2.26}$$

$$\begin{aligned} H_{\mathrm{BdG}}(\boldsymbol{r}) &= \frac{1}{N} \sum_{\boldsymbol{k}} \mathrm{e}^{i\boldsymbol{k}\boldsymbol{r}} H_{\mathrm{BdG}}(\boldsymbol{k}) \\ &= \begin{pmatrix} H_0(\boldsymbol{r}) & -\Delta(\boldsymbol{r}) \\ -\Delta^*(\boldsymbol{r}) & -H_0^*(\boldsymbol{r}) \end{pmatrix} . \end{aligned} \tag{2.27}$$

The Schroedinger equation in real space

$$H_{\mathrm{BdG}}(\boldsymbol{r})\Psi(\boldsymbol{r}) = E\Psi(\boldsymbol{r}) \tag{2.28}$$

is termed the *Bogoliubov-de Gennes equation.* With the BdG equation, the local QP excitations can be studied under a local potential, e.g. a vector potential $H_0 \propto \boldsymbol{A}(\boldsymbol{r})$, inside the superconductor or one can impose a local form of the gap $\Delta(\boldsymbol{r})$ in the vicinity of a defect.

2.2.3 Migdal-Eliashberg Theory

In this section, we want to discuss the mechanism of superconductivity in the presence of a strong-electron phonon interaction. One now has to take a closer look at the electron-electron scattering process via the exchange of a virtual phonon, as shown in the diagrammatic comparison of the interaction vertices for the BCS and Eliashberg theory in Fig. 2.4. Because this process involves time-dependent potentials, it is useful to do so using field operators in the interaction picture and employ the Green's function technique. Without going into much detail of the theory, we want to summarize in simple terms that knowing a system's single-particle Green's function $G(\boldsymbol{r}-\boldsymbol{r}', t-t')$ means knowing its single-particle Hamiltonian and thus the time evolution of physical observables. We already saw that the attractive electron-electron interaction is a dynamical process. Hence, it makes most sense to consider the Fourier-transformed Green's function in momentum and frequency space. For completeness, we want to introduce the Matsubara Green's function in the imaginary frequencies [70]

$$G_\sigma(\boldsymbol{k}, i\omega_n) = -\frac{1}{\sqrt{\beta V}} \int_0^\beta \mathrm{d}\tau \int_{-\infty}^{\infty} \mathrm{d}\boldsymbol{r} \langle \mathcal{T}_\tau \Psi_\sigma(\boldsymbol{r},\tau) \Psi_\sigma^\dagger(0,0) \rangle \mathrm{e}^{i\omega_n \tau - i\boldsymbol{k}\boldsymbol{r}} \tag{2.29}$$

with $\beta = 1/k_B T, \tau = it$ and the fermionic Matsubara frequencies $\omega_n = (2n+1)\pi/\beta$. $\langle .. \rangle$ is the grand canonical average, $\mathcal{T}_\tau$ the time ordering operator and Ψ_σ and $\Psi_\sigma^\dagger$ are annihilation and creation operators for the fermion field with spin σ. The physically relevant retarded Green's function can always be obtained from the Matsubara (thermal) Green's function by analytic continuation of the imaginary frequencies onto the real axis:

$$G^{\mathrm{R}}(\boldsymbol{k}, \omega) = G(\boldsymbol{k}, i\omega_n \to \omega + i0^+). \tag{2.30}$$

In the presence of strong electron-phonon coupling, the electronic excitations and lattice vibrations cannot be treated independently. Instead, the polarisation of the lattice by fluctuations in the electron density exerts a back action on the electronic system leading to electrons and phonons with renormalized dispersion relations. The dressed electron and phonon propagators can be evaluated in a Dyson series of the normal propagators (Green's functions) G_0:

$$G = G_0 + G_0 \Sigma G = G_0 + G_0 \Sigma G_0 + G_0 \Sigma G_0 \Sigma G_0 + ... \tag{2.31}$$

The electronic and bosonic complete self-energies Σ and Π due to the electron-phonon interaction are then defined by the equations

$$\left[G(\boldsymbol{k}, i\omega_n)\right]^{-1} = \left[G_0(\boldsymbol{k}, i\omega_n)\right]^{-1} - \Sigma(\boldsymbol{k}, i\omega_n), \tag{2.32}$$

$$\left[D(\boldsymbol{q}, i\Omega_n)\right]^{-1} = \left[D_0(\boldsymbol{q}, i\Omega_n)\right]^{-1} - \Pi(\boldsymbol{q}, i\Omega_n). \tag{2.33}$$

The phonon Green's function D is defined analogously to Eq. (2.29) with boson field operators expanded in plain waves of momentum $\boldsymbol{q}$ and the bosonic Matsubara frequencies $\Omega_n = 2n\pi/\beta$.

The Dyson equation (2.31) is already complete and can be described by a diagrammatic representation. But the self-energy still consists of a number of connected Feynman diagrams and in the case of electron-phonon interaction, it is useful to sort these

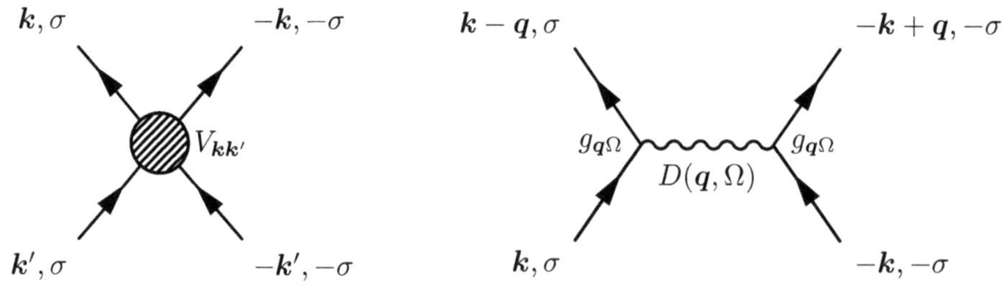

Figure 2.4: Electron-Electron Interaction: Feynman diagrams depicting the attractive electron-electron interaction through the effective four-fermion vertex in BCS theory (left) and through the virtual exchange of a phonon with wavevector $\boldsymbol{q}$ and frequency Ω in Eliashberg theory (right).

diagrams after their weight and write the Dyson equation in the self-consistent Born approximation (SCBA)

$$G = G_0 + G_0 \Sigma G = G_0 + G_0(\Sigma^{(1)} + \Sigma^{(2)} + ..)G. \tag{2.34}$$

This is the form in which we want to make an important approximation. Since the movement of the electrons adiabatically follows the movement of the ions (Born-Oppenheimer theorem) the interaction vertex is not renormalized but taken as the bare electron-phonon vertex. This is *Migdal's theorem* [65]. The neglected terms are of order $\mathcal{O}(\sqrt{m/M})$, with m being the electron and M the ion mass, so only a few percent. This is diagrammatically depicted in Eq. (2.37). Eliashberg used the same argument to cut off the SCBA Dyson series for the electron and phonon propagators after the first fully self-consistent self-energy correction [17]. This is depicted in Eq. (2.35) and (2.36).

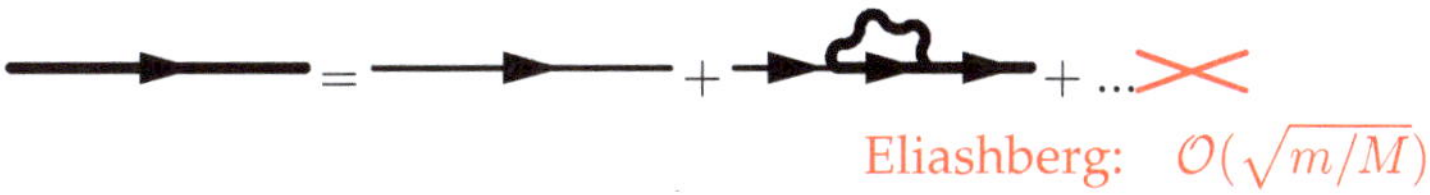

(2.35)

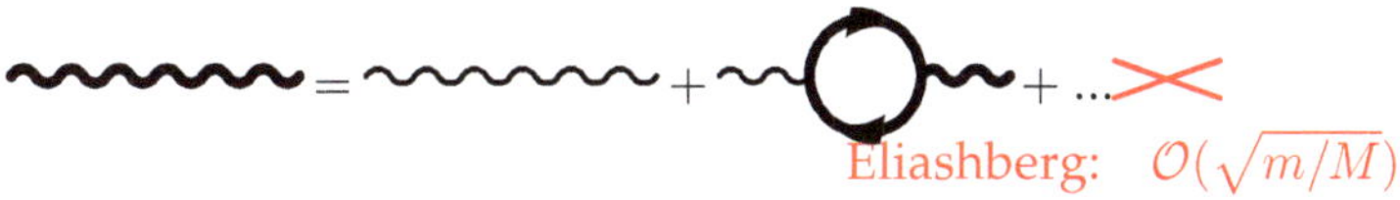

(2.36)

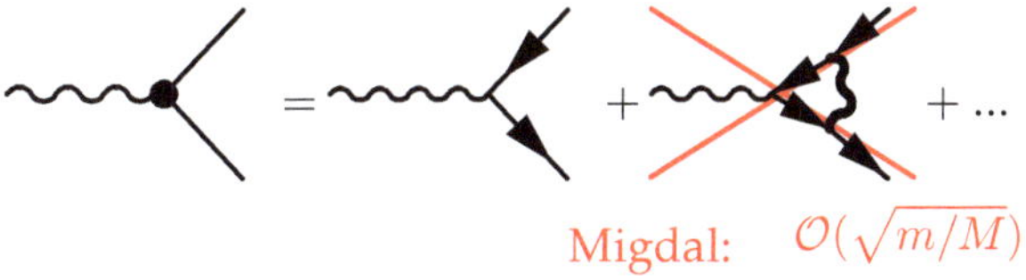

(2.37)

The self-energies are thus given by

$\Sigma \approx \Sigma^{(1)}_{\mathrm{SCBA}} =$ 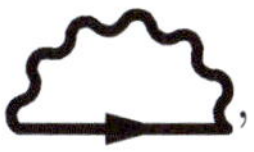, (2.38)

$\Pi \approx \Pi^{(1)}_{\mathrm{SCBA}} =$ . (2.39)

The last thing that is needed before the effective electron-electron interaction can be analyzed is the extension of the electronic Green's function to electron-hole space (Nambu spinor representation). It is then given by the 2×2 matrix [70]

$$\hat{G}(\boldsymbol{k},\tau) = -\begin{pmatrix} \langle T_\tau c_{\boldsymbol{k}\uparrow}(\tau) c^\dagger_{\boldsymbol{k}\uparrow}(0) & \langle T_\tau c_{\boldsymbol{k}\uparrow}(\tau) c_{-\boldsymbol{k}\downarrow}(0)\rangle \\ \langle T_\tau c^\dagger_{-\boldsymbol{k}\downarrow}(\tau) c^\dagger_{\boldsymbol{k}\uparrow}(0)\rangle & \langle T_\tau c_{-\boldsymbol{k}\downarrow}(\tau) c^\dagger_{-\boldsymbol{k}\downarrow}(0)\rangle \end{pmatrix} := \begin{pmatrix} G_{\boldsymbol{k}} & F_{\boldsymbol{k}} \\ \overline{F}_{\boldsymbol{k}} & -G_{-\boldsymbol{k}} \end{pmatrix} \tag{2.40}$$

where the diagonal entries describe the normal propagators for spin-up electrons and spin-down holes and the off-diagonal entries are Gorkov's anomalous propagators. Just like in the BCS theory these anomalous operator averages describe the pairing mechanism. They naturally appear in the evaluation of the right diagram in Fig. 2.4 where two electrons scatter by exchange of a virtual phonon and in the boson self-energy. The two equations for the fermion and boson self-energies, Eq. (2.38) and (2.39), build, together with Eq. (2.35) and (2.36), a set of coupled integral equations that can be solved numerically. These are the *Eliashberg equations*. For their explicit form, which is lengthy and physically not too insightful, the reader is referred to standard literature.

An essential ingredient of the Eliashberg equations is the defnition of the *Eliashberg function* [70]

$$\alpha^2 F(\Omega) = -\frac{1}{V^2 \nu_{\mathrm{n}}^F} \sum_{\boldsymbol{q}=\boldsymbol{k}-\boldsymbol{k}'} \delta(\epsilon_{\boldsymbol{k}})\delta(\epsilon_{\boldsymbol{k}'}) \frac{|g_{\boldsymbol{q}\Omega}|^2}{\pi} \mathbf{Im} D^{\mathrm{R}}(\boldsymbol{q},\Omega)\Theta(\Omega). \tag{2.41}$$

It is a momentum-integrated phonon DOS that electrons, which scatter between points of the Fermi surface, see. It is often also called the boson spectral function or bosonic glue in order to generalize it to the exchange of a different boson. In general Eliashberg theory, the gap becomes energy dependent $\Delta \to \Delta(\epsilon)$ and is mainly determined through the Eliashberg function. In a conventional s-wave superconductor a sharp resonance in the Eliashberg function at Ω^* leads to additional fine structure in the electronic DOS

$$\nu(\epsilon) = \nu_{\mathrm{n}}^F \mathbf{Re} \frac{|\epsilon|}{\sqrt{\epsilon^2 - \Delta(\epsilon)^2}} \tag{2.42}$$

at energies $\epsilon = \Delta_0 + n \cdot \Omega^*$, as was demonstrated both theoretically [70] and in tunnelling experiments on Pb [71–74]. By defining the dimensionless electron-phonon coupling constant [18]

$$\lambda = 2 \int \mathrm{d}\Omega \frac{\alpha^2 F(\Omega)}{\Omega}, \tag{2.43}$$

the consequence of the coupling strength can be understood in the BCS limit $\lambda \ll 1$. Electrons effectively gain a mass proportional to λ and the effective scattering potential V in Eq. (2.14) is substituted by λ, which nicely illustrates how a strong electron-phonon coupling leads to a stronger electron-electron attraction. Strong-coupling superconductors, like Pb, have larger $\Delta(0)/T_c$ ratios than the BCS value (Eq. (2.23)).

2.3 Multiband Superconductivity

The simultaneous existence of superconductivity in more than one electronic band is not surprising per se. After all, many superconducting materials have more than one electronic band crossing the Fermi level. The interplay of several superconducting bands has, however, been mostly studied in the model two-band superconductor MgB_2 [36–38] and iron-pnictide superconductors [75–77]. For elemental superconductors, supposed evidence for more than one superconducting gap was found in the heat capacity of Nb, Ta and V in the 1960s [78]. Due to contradicting experimental results in the following years [79–81], however, the influence of multiband effects in these materials has remained a controversial topic. In recent years, modern theoretical methods and higher resolution in energy gap probing experiments paved the way for the discovery of two-band superconductivity in elemental Pb [39] after its prior theoretical prediction [82].

In a simple picture, the system of a two-band superconductor can be described by a model Hamiltonian that incorporates interband and intraband pairing, as well as interband scattering [83]:

$$\begin{aligned} H = &\sum_{\boldsymbol{k},\sigma} \epsilon_{a,\boldsymbol{k}} c^{\dagger}_{a,\boldsymbol{k},\sigma} c_{a,\boldsymbol{k},\sigma} + \epsilon_{b,\boldsymbol{k}} c^{\dagger}_{b,\boldsymbol{k},\sigma} c_{b,\boldsymbol{k},\sigma} \\ &+ \sum_{\alpha,\beta,\boldsymbol{k}} \Delta_{\alpha\beta} c^{\dagger}_{\alpha,-\boldsymbol{k},\uparrow} c^{\dagger}_{\beta,\boldsymbol{k},\downarrow} + \text{h.c.} \\ &+ \sum_{\boldsymbol{k},\sigma} \Gamma c^{\dagger}_{a,\boldsymbol{k},\sigma} c_{b,\boldsymbol{k},\sigma} + \text{h.c.}\,. \end{aligned} \tag{2.44}$$

$\epsilon_{\alpha,\boldsymbol{k}}$ denotes the quasiparticle dispersion in band α, $c^{\dagger}_{\alpha,\boldsymbol{k},\sigma}$ is the creation operator for a QP with momentum $\boldsymbol{k}$ and spin σ in the band α, $\Delta_{\alpha,\beta}$ is the superconducting gap parameter for inter- or intraband pairing and Γ describes the QP scattering amplitude between bands a and b. A large interband scattering amplitude $\Gamma \lesssim \Delta$ can lower the transition temperature T_c in a linear manner [84], a small $\Gamma \ll \Delta$ enables one to effectively describe the multiband superconductor in a single-band model for each band. Consequently, the DOS may then be written as a superposition of contributions from each individual band α, both in the normal and superconducting state:

$$\nu_{s(n)}(\omega) = \sum_{\alpha} \nu_{s(n),\alpha}(\omega). \tag{2.45}$$

In a tunnelling experiment, the individual band DOS may be probed with different spectral weight due to the momentum and energy dependence of the tunnelling matrix element. On a two-band superconductor this is notably seen as a discrepancy in the differential conductance at the quasi-particle coherence peaks [38, 39].

In order to determine how large the interband scattering amplitude is, one can look for hybridization-induced high energy gaps in the superconducting state, that have been predicted by Komendová et al. [85]. These gaps are a consequence of avoided band crossings of the hybridized bands at higher energies, hence their energetic position depends on the normal state dispersion. This exotic odd-ω pairing state can only be formed if $\Delta_1 \neq \Delta_2$ and $\Gamma > 0$. The higher Γ, the larger the width of the hybridization-induced gaps, that appear symmetrically around the Fermi energy in a conductance spectrum. On top of that, the translational symmetry breaking of a vortex in an s-wave superconductor is supposed to lead to an odd-frequency pairing mechanism inside the vortex core [86]. Another theoretically predicted consequence of a large interband coupling is a coherence length and gap parameter matching. As was shown by Ichioka et al., for growing interband coupling, the vortex core radii ratio $\xi_1^{(c)}/\xi_2^{(c)} \rightarrow 1$ and maximum gap sizes $\Delta_2/\Delta_1 \rightarrow 1$ [87].

2.4 Mott Insulators

It may sound paradoxical to talk about insulators when dealing with superconductivity, but for reasons that are still not perfectly understood, materials with enhanced critical temperatures very often feature a MIT at temperatures larger than T_c. This includes organic superconductors, fullerenes, bismuthates and the two classes that shall be addressed in more detail in the following: Granular superconductors and cuprates [88, p. 63]. In these classes of superconductors, the MIT is of Mott type. This means that a transition from the metallic to insulating state is caused by a charge carrier density that is so low that Thomas Fermi screening becomes very weak and electrons are localized on individual atoms due to strong on-site Coulomb repulsion. For a better understanding, it is instructive to first consider a metal crystal like copper where the conduction electrons can be thought of as a Fermi gas, so non-interacting particles, due to the effective Thomas Fermi screening. In a gedanken experiment proposed by Mott, a Mott type MIT could be initiated in such a crystal by just pulling the ions further and further apart and by doing so effectively reducing the charge carrier density through the increase in crystal volume[2]. The Thomas Fermi

[2] This is of course under the assumption that the crystal does not undergo a structural transition and does not gain a large number of defects in the process of its expansion.

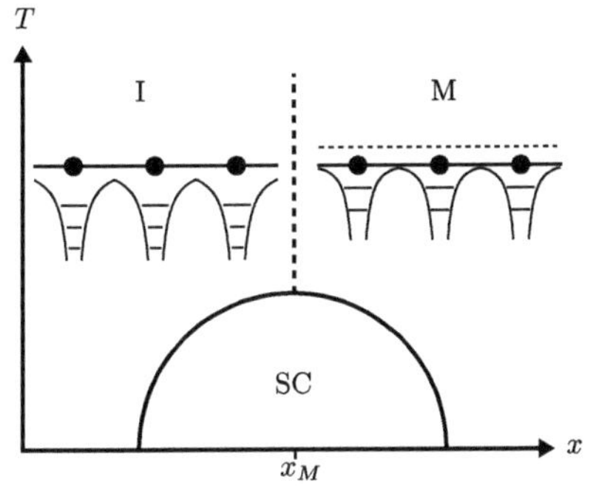

Figure 2.5: Mott MIT: Schematic of a generic phase diagram of superconductors with a Mott MIT. As the control parameter x, that regulates the charge carrier density, falls below x_M, the material undergoes a discontinuous phase transition from metal to Mott insulator. Upon decreasing T below T_c in a region of x around x_M, the material undergoes a continuous phase transition to the superconducting state.

screening length r_{TF} in the Fermi gas, which was the initial state, is becoming larger for lower charge concentrations like $r_{\mathrm{TF}}^2 \propto n^{-1/3}$. As soon as r_{TF} becomes much larger than the interatomic distance a, the Thomas Fermi screening is ineffective. The potential wells of neighbouring ions overlap and only bound states are formed. The strong electron-electron Coulomb repulsion at each atomic site renders electronic wavefunctions localized. In this insulating state, strong electron-electron interactions are dictating the materials electronic properties and as a consequence, a theoretical description in terms of single-particle states is not possible. Instead, a theoretical understanding of the ground state is usually gained in a local description of pair-wise interactions, like in the Hubbard model with the Hamiltonian:

$$H = -t_{ij} \sum_{\langle ij \rangle, \sigma} (c_{i\sigma}^{\dagger} c_{j\sigma} + c_{j\sigma}^{\dagger} c_{i\sigma}) + U \sum_{i} n_{i\uparrow} n_{i\downarrow}. \tag{2.46}$$

Here, $c_{i\sigma}^{(\dagger)}$ are the annihilation (creation) operators for a fermion with spin σ at lattice site i and $n = c^{\dagger}c$ are the fermionic density operators. The first term contains the tight-binding band structure via a hopping term t_{ij} between neighbouring sites $\langle ij \rangle$. The second term describes the on-site Coulomb repulsion of two electrons or holes via the potential U. From this Hamiltonian, it becomes obvious how a critical point can be reached when the decrease in energy due to the kinetic term is overcompensated by an increase in potential energy due to the on-site Coulomb repulsion.

In Fig. 2.5, a highly simplified, schematic phase diagram is shown for superconductors with a Mott MIT. When the control parameter x that regulates the charge carrier density falls below x_M, the material undergoes a discontinuous phase transition from metal to Mott insulator, indicated by the dashed line. Upon decreasing T below T_c in a region of x around x_M the material undergoes a continuous phase transition to the superconducting state, indicated by the continuous line. The small cartoon depicts how an increased screening length leads to the formation of more localized states. Granular superconductors and cuprates are two prime examples in which the superconducting phase not only exists in close proximity to a Mott MIT, it also exhibits some of its highest transition temperatures in the vicinity of it [88, p. 64].

2.4.1 Granular Superconductors

Granular Superconductors are materials in which small metallic grains are separated by an insulating barrier. The control parameter that regulates the charge carrier density is the intergrain resistance, so the resistivity or thickness of the barrier. Here, the Mott insulator is characterized by a localization of electronic wavefunctions on single grains, as soon as the intergrain coupling becomes too weak. In a sense, the granular system is a mesoscopic Mott insulator [88, p. 68], because nanosized grains play the role of single atoms. A pair of metallic grains, separated by the insulating buffer, represents a normal metal tunnel junction with resistance R and capacitance C. The electrostatic energy of such a junction can only change in quantized decrements of $E_c = q^2/2C$, corresponding to the tunnelling of a single charge q. As a consequence, a tunnelling current can only flow above a threshold voltage of

$$|U| \geq U_{\mathrm{CB}} = \frac{Q}{2C}. \tag{2.47}$$

This constraint is called the *Coulomb blockade*. It can be overcome by thermal or quantum fluctuations. Hence, in order to induce metallic behaviour, the thermal energy $k_B T$ has to be larger than E_c or the time scale on which a single charge transfer occurs through quantum fluctuations $\tau = RC$ has to become sufficiently short, i.e. $E_c \leq h/\tau$ which requires $R \leq R_K$, where $R_K = h/e^2$ is the von Klitzing constant [89, 90].

In the superconducting state, the metallic grains become superconducting and the junctions are effective Josephson junctions with Josephson coupling E_J. The tunnelling charge carriers now have the charge $q = 2e$. Since the number of Cooper pairs on each grain N and the phase of the BCS wavefunction φ are conjugate variables, i.e. $[N, \varphi] = i$, the charge per grain $Q = N(2e)$ and the phase cannot be measured independently. In the case of extremely decoupled grains, $E_c \ll E_J$, the junction is in the Coulomb blockade regime and experiences large fluctuations of the superconducting phase. In the case of extremely well coupled grains, $E_J \gg E_c$, the junction is in the phase blockade regime, which means the superconducting phase is locked but charge fluctuations are large.

2.4.2 Cuprates

In copper-oxide based superconductors, the cuprates, the Mott criterium is evoked on the atomic scale. There, a transition from the antiferromagnetic Mott insulator to a metallic phase is initiated through the introduction of additional charge carriers via doping. A more detailed discussion of the phase diagram of the cuprates is presented in Sec. 4.1.

2.5 Unconventional Superconductivity

A superconductor is classified as unconventional if the electron pairing mechanism is not mediated by electron-phonon interactions or additional symmetries besides the global $U(1)$ symmetry are spontaneously broken for $T < T_c$ [91, p. 3]. This unconventional pairing mechanism can lead to an anisotropic gap function $\Delta_{\boldsymbol{k}}$ that is not explained by the BCS mean-field theory. In order to explain it, it is necessary to include strong electron correlations and local interactions. Especially for the technologically interesting high-temperature copper oxide superconductors, a magnetic order is competing with a superconducting phase. In these systems, the conduction and magnetic interactions are confined to the copper oxide planes and are very sensitive to doping, so the charge carrier density. Despite the absence of a complete microscopic theory that explains the unconventional electronic pairing in these systems, considerable progress has been made in the framework of a spin-fluctuation mediated pairing mechanism. This mechanism can explain the close proximity of superconductivity and magnetism as well as the resulting order parameter symmetry in the copper oxide superconductors [92].

2.5.1 Order Parameter Symmetry

The classification of superconductors into groups based on the symmetry of their pairing amplitude proves to be a powerful concept to make good physical predictions without tedious calculations. We recall that the pairing amplitude is given by

$$\Delta_{\sigma_1\sigma_2}(\boldsymbol{k}) = -\frac{1}{N}\sum_{\boldsymbol{k}'} V_{\boldsymbol{k}\boldsymbol{k}'\sigma_1\sigma_2}\langle c_{-\boldsymbol{k}'\sigma_1} c_{\boldsymbol{k}'\sigma_2}\rangle \tag{2.48}$$

and is determined in the self-consistent Eq. (2.14). In this eigenvalue problem of T_c, $\Delta_{\sigma_1\sigma_2}(\boldsymbol{k})$ are the eigenfunctions which need to have certain symmetry properties based on their definition. The phase transition from the normal conducting to superconducting state is a second-order phase transition and therefore tightly connected with the reduction of the underlying symmetry of the system's ground state. If the full symmetry group G describes the system's normal state, then the superconducting state belongs to a subgroup H of G [56]

$$H \subset G, \quad T < T_c. \tag{2.49}$$

An important implication of spontaneous symmetry breaking is that the Hamiltonian that describes the time-evolution of the system is still invariant under all symmetry operations of G. Only the ground state has a reduced symmetry and is not invariant under symmetry operation corresponding to the broken part of G. Starting from the simplest consideration of a Fermi liquid, in which the interaction from Eq. (2.48) is isotropic in real space and momentum space, we only need to account for symmetry of the fermionic wavefunctions, i.e. respect the Pauli principle. The full symmetry group G would look as follows:

$$G = \underbrace{SU(2)}_{\text{spin-space}} \times \underbrace{SO(3)}_{\text{k-space}} \times \underbrace{U(1)}_{\text{phase}} \times \underbrace{T}_{\text{time-reversal}} \tag{2.50}$$

First, we want to use the fact that eigenfunctions of the system are built from a basis of an irreducible representation (irrep) of the symmetry group. A representation of the group H would be

$$\mathcal{D}_H = \mathcal{D}_l \otimes \mathcal{D}_s \otimes \mathcal{D}_T \tag{2.51}$$

where $\mathcal{D}_l$ is an irrep of momentum space and $\mathcal{D}_s$ and $\mathcal{D}_T$ denote representations of spin-space and the time-reversal symmetry [93]. Analogously to the hydrogen atom, we can use the spherical harmonics $Y_{lm}(\boldsymbol{k})$ as the irrep of the $SO(3)$ in momentum space and since our fermions have a spin of $1/2$ we can use the spin functions χ_{sm_s}, which are the eigenvectors of the 2D Hermitian Pauli matrices. The eigenfunctions to the eigenvalues $T_c(l, s)$ can then be written as [93]

$$\Delta_{l,s}(\boldsymbol{k}) = \begin{cases} \sum\limits_{m} c_m Y_{lm}(\boldsymbol{k}) \chi_{00} & \text{Singlet}, l = \text{even} \\ \sum\limits_{m,m_s} c_{mm_s} Y_{lm}(\boldsymbol{k}) \chi_{1m_s} & \text{Triplet}, l = \text{odd} \end{cases} \tag{2.52}$$

with complex scalar coefficients c_m and c_{m,m_s}. Generally, the highest eigenvalue $T_c(l, s)$ and its corresponding eigenfunction dominate the expansion of the order parameter. However, we can already understand the order parameter composition of eigenfunctions with different orbital or spin character from this very simple example. Of special importance is the transformation behaviour of the order parameter under parity operation $\mathcal{P}$. We classify the superconductors according to their parity eigenvalue $\mathcal{P}\Delta(\boldsymbol{k}) = \Delta(-\boldsymbol{k}) = (-1)^l \Delta(\boldsymbol{k})$ into s-wave (l=0), p-wave (l=1), d-wave (l=2) and so on.

Table 2.1: Residual Groups: Spin-Singlet Pairing States with Even Parity in D_{4h} (a) and D_{2h} (b) [94]

(a)

Irrep*	Residual group†	Basis functions
A_{1g}	$D_{4h} \times T$	$1, x^2+y^2, z^2$
A_{2g}	$D_4(C_4) \times C_i \times T$	$xy(x^2-y^2)$
B_{1g}	$D_4(D_2) \times C_i \times T$	x^2-y^2
B_{2g}	$D_4(D_2') \times C_i \times T$	xy
E_g	$D_2(C_2') \times C_i \times T$	xz
	$D_2(C_2'') \times C_i \times T$	$(x+y)z$
	$D_4(E) \times C_i$	$(x+iy)z$

(b)

Irrep*	Residual group†	Basis functions
A_{1g}	$D_{2h} \times T$	$1,x^2,y^2,z^2$
B_{1g}	$D_2(C_2^z) \times C_i \times T$	xy
B_{2g}	$D_2(C_2^y) \times C_i \times T$	xz
B_{3g}	$D_2(C_2^x) \times C_i \times T$	yz

*We use Mulliken symbols., †All residual symmetry groups also contain the $SU(2)$ of spin-space.

So far we have neglected spin-orbit interaction and the crystal field. Strong spin-orbit interaction will mix eigenstates of l and s, but we can save our basis by using the total angular momentum number j and an effective pseudo-spin α [93]. Introducing the crystal field will reduce the symmetry group in k-space from a continuous rotation group $SO(3)$ to the discrete crystallographic point group G_c. Irreps of the crystallographic point groups are tabulated in numerous review articles on this subject [93, 94]. Since our focus will be on the cuprates, we will concentrate on the orthorhombic point group D_{2h} and compare it to the irreps of the point group D_{4h}, the point group of a tetragonal crystal. The irreps of these two point groups for singlet-pairing (even parity) are shown in Tab. 2.1. A BCS superconductor can only be represented in the A_{1g}, the only irrep where the residual group is identical to the point group of the normal state except for the global $U(1)$ symmetry. No additional symmetry is broken. The basis function for A_{1g} is 1. This means that the gap parameter can be written as a constant, which is equivalent to the conventional s-wave pairing state. In unconventional superconductors, however, an additional

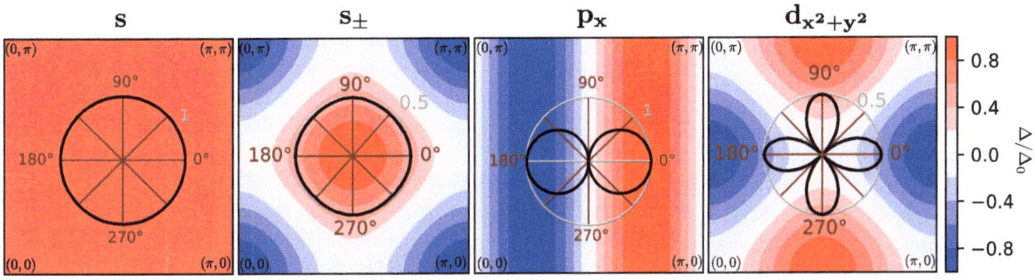

Figure 2.6: Pairing Symmetries: 2D sketch of several (an)isotropic gap parameter functions $\Delta(\boldsymbol{k})$ inside the first Brillouin zone. The inner radial plot shows the polar dependence of Δ for a fixed $|\boldsymbol{k}|$ (isotropic Fermi surface).

symmetry is spontaneously broken and in most cases this symmetry breaking occurs in $\boldsymbol{k}$-space. The symmetry of the pairing state thus has a lower symmetry in momentum space than the crystal lattice. According to the basis functions of a certain irrep from Tab. 2.1, the gap function for a certain pair state $\Delta(\boldsymbol{k})$ can be expanded in harmonics of $\cos(k_i)^3$ where k_i with $i = x, y, z$ are the k-vectors in three orthogonal directions inside the Brillouin zone [95]. This is analagous to the atomic orbitals, where linear combinations of spherical harmonics are used to obtain real wavefunctions that are often denoted in the cartesian basis, e.g. a d_{z^2} or $d_{x^2-y^2}$ orbital. For the s-wave state (A_{1g}) of the tetragonal crystal with point group D_{4h}, one obtains [56, 95]

$$\Delta_s(\boldsymbol{k}) = \Delta_s^{(0)} + \Delta_s^{(1)}(\cos k_x + \cos k_y) + \Delta_s^{(2)} \cos k_z + ... \tag{2.53}$$

and for the $d_{x^2-y^2}$-wave state (B_{1g})

$$\Delta_{d_{x^2-y^2}}(\boldsymbol{k}) = \Delta^0_{d_{x^2-y^2}}(\cos k_x - \cos k_y) + ... \quad . \tag{2.54}$$

Apart from these pure pairing states, it is also possible to construct mixed pair states by adding the gap parameter times i of one representation to the gap parameter of another representation, provided that both representations are 1D [95]. An example for D_{4h} would be

$$\Delta_s(\boldsymbol{k}) + i\Delta_{d_{x^2-y^2}}(\boldsymbol{k}). \tag{2.55}$$

It should be noted that these mixed states break time-reversal symmetry and result in two first-order phase transitions instead of one second-order phase transition [56]. Remarkable in the point group D_{2h} is that the s- and $d_{x^2-y^2}$-pairing state belong to the same irrep A_{1g}. A simple superposition of the two is therefore allowed without building a time-reversal symmetry breaking mixed state or requiring two unrelated phase transitions. A more detailed discussion on the pairing symmetry in the two cuprates studied in this work follows in Sec. 4.1.2.

2.5.2 Macroscopic and Microscopic Description of Magnetic Response in a Metal

In order to understand how Cooper pair formation can be mediated by spin-fluctuations, it is first important to understand how the spins of itinerant electrons collectively interact with an external magnetic field. Even though the microscopic interactions may be very intricate, from a macroscopic viewpoint, the magnetic

[3] Using Taylor-series expansion of the cosine and sine functions make it obvious why e.g. $k_x^2 - k_y^2 \sim \cos k_x - \cos k_y \approx (1 + k_x^2/2) - (1 + k_y^2/2)$.

properties of such a system can be described by one function: the magnetic susceptibility $\chi = \partial \boldsymbol{M}/\partial \boldsymbol{H}$. It is the response function of the system to an external magnetic field. Since the perturbation may be time-dependent, it is favourable to study the Fourier-transformed magnetic susceptibility, i.e. how the system reacts to an external field of a certain frequency ω and wave-vector $\boldsymbol{q}$:

$$\chi(\boldsymbol{q},\omega) = \int\int \chi(\boldsymbol{r},t)\mathrm{e}^{-i(\boldsymbol{q}\boldsymbol{r}-\omega t)}\mathrm{d}\boldsymbol{r}\mathrm{d}t. \tag{2.56}$$

Since the magnetic susceptibility has to obey the principle of causality, its real and imaginary part are not independent, but related to each other through Kramers-Kronig relations. In other words, it is sufficient to either perform a diffractive scattering experiment to probe the real part of χ or an inelastic scattering experiment to probe the imaginary part responsible for damping.

Microscopically, the magnetic moment operator $\hat{\boldsymbol{M}}(\boldsymbol{r})$ per unit volume is the sum over all electron magnetic moment operators $\hat{\boldsymbol{m}}_i$

$$\hat{\boldsymbol{M}}(\boldsymbol{r}) = \sum_i \hat{\boldsymbol{m}}_i \delta(\boldsymbol{r}-\boldsymbol{r}_i). \tag{2.57}$$

Solving the time-dependent Schroedinger equation for an adiabatically turned on external field and using statistical methods, one is able to relate microscopic and macroscopic description through a magnetization (or spin) correlation function [96, p. 15]

$$\chi_{\mu\nu}(\boldsymbol{q},\omega) = \frac{i}{\hbar}\mu_0 \int_0^\infty \langle[\hat{M}_\mu(\boldsymbol{q},t),\hat{M}_\nu(-\boldsymbol{q},0)]\rangle \mathrm{e}^{i\omega t}\mathrm{d}t, \tag{2.58}$$

where $\langle\hat{A}\rangle = \mathrm{Tr}(\rho\hat{A})$ is the quantum-statistical average using the density matrix ρ and $[\hat{A},\hat{B}] = \hat{A}\hat{B} - \hat{B}\hat{A}$ is the operator commutator. Introducing spin fluctuations by deviation from the mean value $\delta\hat{\boldsymbol{M}} = \hat{\boldsymbol{M}} - \langle\hat{\boldsymbol{M}}\rangle$ and using the fluctuation-dissipation theorem the longitudinal dynamic structure factor $S(\boldsymbol{q},\omega)$ can be directly linked to the imaginary part of the magnetic susceptibility [96, p. 17]:

$$\begin{aligned} S_{\mu\mu}(\boldsymbol{q},\omega) &= \frac{\mu_0}{\hbar}\int_{-\infty}^{\infty} \langle\delta\hat{M}_\mu(\boldsymbol{q},t)\delta\hat{M}_\mu(-\boldsymbol{q},0)\rangle \mathrm{e}^{i\omega t}\mathrm{d}t \\ &= \frac{2}{1-\mathrm{e}^{-\hbar\omega/k_BT}}\mathrm{Im}\chi_{\mu\mu}(\boldsymbol{q},\omega). \end{aligned} \tag{2.59}$$

This structure factor is directly accessible in an INS experiment. The spin correlation function and hence the magnetic susceptibility in Eq. (2.58) diverge, when the system undergoes a phase transition to a magnetically ordered state $\chi(\boldsymbol{q} \to \boldsymbol{Q}) \to \infty$. The wave vector at which the system is most prone to do that (highest transition temperature) is called the magnetic ordering wave vector $\boldsymbol{Q}$. Energy dissipated into the electronic system through a magnetic interaction (cf. Eq. (2.59)) can be taken up in two ways: It can produce a spin-wave, a collective excitation of the electronic spins with momentum $\boldsymbol{q}$ and energy $\hbar\omega$ or produce Stoner excitations. The former can be described by a bosonic excitation (a magnon) with a single-particle dispersion $E(\boldsymbol{q},\omega)$ whereas the latter form a continuum, typically in the long wave-vector and high-energy sector, where they can be thought of as local spin flips.

2.5.3 Spin-Fermion Model

Even without an external field, the itinerant electrons feel an effective magnetic field at finite temperature due to the constant temperature driven fluctuations of magnetic moments. Even though the valence electrons try to quickly screen the local spin polarizations, the process is not immediate and a retarded interaction between electrons and the background magnetic field is the consequence. Especially hot electrons, that carry enough energy and momentum to emit a magnon, can effectively attract another electron further away that feels the polarization of the environment and ultimately resolves it by absorbing the magnon. At low temperatures, this process happens only virtually (off shell) involving electrons and holes in a small energy window around the Fermi edge. The corresponding Feynman diagram to this coupling mechanism is shown in Fig. 2.7. The spin-fermion model aims to provide an effective Hamiltonian that describes the interaction between low-energy electrons and holes and collective spin-excitations. A simple Hamiltonian for this low-energy regime, assuming a short-range Hubbard-type interaction between fermions and spin fluctuations, reads [22, 97]

$$\mathcal{H} = \sum_{\boldsymbol{k},\alpha} \boldsymbol{v}_{\boldsymbol{k}}(\boldsymbol{k}-\boldsymbol{k}_F)c^\dagger_{\boldsymbol{k},\alpha}c_{\boldsymbol{k},\alpha} + \sum_{\boldsymbol{q}} \chi_0^{-1}(\boldsymbol{q})\boldsymbol{S}_{\boldsymbol{q}}\boldsymbol{S}_{-\boldsymbol{q}} + g \sum_{\boldsymbol{k},\boldsymbol{q},\alpha,\beta} c^\dagger_{\boldsymbol{k}+\boldsymbol{q},\alpha}\sigma_{\alpha,\beta}c_{\boldsymbol{k},\beta}\boldsymbol{S}_{-\boldsymbol{q}}. \tag{2.60}$$

Here $c^\dagger_{\boldsymbol{k},\alpha}$ denotes a fermion creation operator, σ_i are the Pauli matrices and $\boldsymbol{S}_{\boldsymbol{q}}$ describe the magnon field. g is the coupling constant between the fermionic and bosonic degrees of freedom and $\chi_0^{-1}(\boldsymbol{q}) = (\xi^{-2} + (\boldsymbol{q}-\boldsymbol{Q})^2)/\chi_0$ is the static part of the spin susceptibility. ξ is the spin correlation length and $\boldsymbol{Q}$ the magnetic ordering vector. The static part of the susceptibility has Ornstein-Zernike form, which captures the short-range interaction as its Fourier transform has the form of a Yukawa potential.

A simple separation between fermionic and bosonic degrees of freedom (which Eq. (2.60) implies), like in the case of electron-phonon interaction, is generally not possible because Migdal's theorem does not apply. Vertex corrections are suppressed by the order $\mathcal{O}(v_s/v_F)$. The major difference compared to the phonon case is that

v_s now describes the spin velocity instead of the sound velocity, which is of the order of the Fermi velocity v_F because the spin fluctuations stem from the fermions themselves. Consequently, not only higher order vertex corrections but also a complete renormalization of the bosonic propagator due to particle-hole bubble diagrams are necessary over a wide range of frequencies [22]. The renormalization of fermionic propagators, spin-polarization operators and vertices follows the diagrammatic series from Eq. (2.35), (2.36) and (2.37) for the phonon exchange, but now higher-order diagrams, which were discarded due to Migdal's theorem before, have to be respected.

Conceptually, one could now proceed analogously to the Migdal-Eliashberg theory and use single fermion propagators that are fully renormalized by the fermionic self-energy $\Sigma(\boldsymbol{k}, \Omega)$ (first order and higher order diagrams in Eq. (2.35)), single boson propagators that are fully renormalized by the bosonic self-energy $\Pi(\boldsymbol{q}, \omega)$ (first order and higher order diagrams in Eq. (2.36)) and the fully renormalized coupling constant $\tilde{g} = g + \Delta g$ (first order and higher order diagrams in Eq. (2.37)). What makes this problem so difficult is the fact that the vertex corrections Δg contain the full fermionic and bosonic propagators. This means that all vertex corrections and self energies have to be solved self-consistently which is a near impossible task. So it is mandatory to look for reasonable justifications to terminate the corrections. This is especially the case for a strong-coupling superconductor.

For the cuprate superconductors, there is a good justification for the neglect of many high-order terms called the hot-spot theory. One makes use of the fact, that the bare spin susceptibility in these systems peaks around the antiferromagnetic ordering vector (AFV) $\boldsymbol{Q}$ and that such a AFV connects different parts of the Fermi surface (hot spots), such that the electron coupling mechanism is also dominated by the spin-fluctuations at $\boldsymbol{Q}$. In the normal state above T_c the full spin susceptibility [22]

$$\chi(\boldsymbol{q}, \omega) = \frac{\chi_0 \xi^2}{1 + \xi^2 (\boldsymbol{q} - \boldsymbol{Q})^2 - \Pi(\boldsymbol{q}, \omega)} \tag{2.61}$$

is then mainly determined by $\Pi(\boldsymbol{Q}, \omega) = i\omega/\omega_{\text{sf}}$, where ω_{sf} is the characteristic energy scale of the boson and a temperature dependent, free parameter. At $\boldsymbol{q} \approx \boldsymbol{Q}$, the spin-fluctuation dynamics are then governed by slow modes [98] which means Migdal's theorem can be applied in this region (no vertex corrections). Additionally,

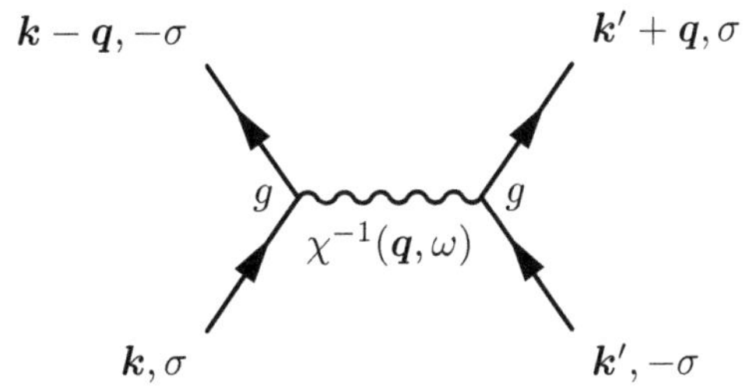

Figure 2.7: (Para)magnon Exchange: Feynman diagram depicting the attractive electron-electron interaction through virtual exchange of a (para)magnon.

it turns out that close to the hot spots, higher order corrections to the fermion and boson propagator only diverge logarithmically in the coupling constant, so a renormalization is possible and the expansion can be cut off after few terms. Now the coupled integral equations can be solved self-consistently, just like in the Eliashberg theory. This is shown in detail in Ref. [21, 70]. The self-energies of the anomalous Green's function $\Phi(\boldsymbol{k},\Omega)$ and normal Green's function $\Sigma(\boldsymbol{k},\Omega)$ can lead to a nonzero gap function $\Delta(\boldsymbol{k},\Omega)$ below T_c, which is now sign changing for related hot spots $\Delta(\boldsymbol{k}+\boldsymbol{Q}) = -\Delta(\boldsymbol{k})$. The role of the Eliashberg function is now played by $g^2\chi''(\omega)$.

The dimensionless, integrated spin spectrum is given by [28]

$$\chi''(\omega) = -3\nu_S^0 \int \mathrm{d}^d q \mathrm{Im}\chi(\boldsymbol{q},\omega)/\pi, \tag{2.62}$$

where ν_S^0 denotes the normal state DOS of the superconductor. The integrated spin spectrum plays the role of the bosonic excitation spectrum in tunnelling experiments as we will see in Sec. 4.2. P. Hlobil and J. Schmalian calculated this bosonic spectrum and related functions self-consistently in the spin-fermion model (see Ref. 2.8). Fig. 2.8 shows their results. They present the imaginary part of the spin susceptibility at the AFV and the spin spectrum integrated over the 2D Fermi surface of an iron-pnictide superconductor with a full gap ($s_\pm$-symmetry). In the normal state, the spin susceptibility shows the expected behaviour for overdamped spin-fluctuations. In the superconducting state the spin susceptibility is drastically renormalized: The spin spectrum at the AFV develops a spin gap for $\omega < \omega_{\mathrm{res}} < 2\Delta$. While a full spin gap is present for $\omega < \Delta$ due to the gapped fermionic degrees of freedom in the superconductor, the break-up of a Cooper pair requires an energy of $\omega > 2\Delta$. In the intermediate region $\Delta < \omega < 2\Delta$ a spin resonance evolves at ω_{res} that couples to the electron-like quasiparticles, creating a peak in the DOS at $\Delta + \omega_{\mathrm{res}}$. The integrated spin spectrum approaches the value of the normal state for $\omega \sim 5\Delta$. One should

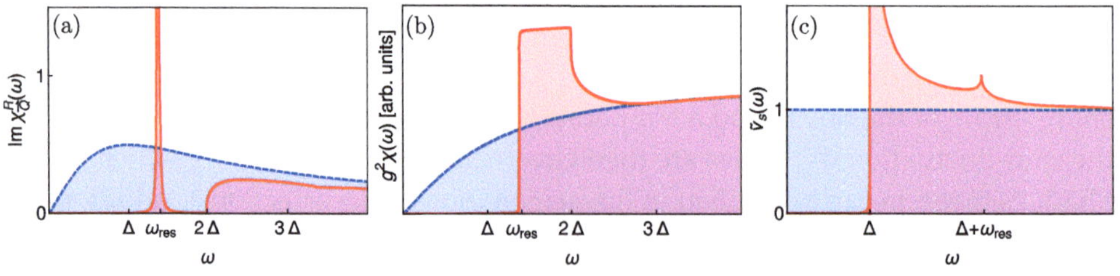

Figure 2.8: Spin-Fermion Model's Spin Resonance: Calculated spectra in the spin-fermion model for the normal state (blue) and superconducting state (red): (a) Spin spectrum at the antiferromagnetic ordering vector (AFV) showing a spin resonance at ω_{res} below T_c. (b) Integrated spin spectrum over the 2D Brillouin zone of a fully gapped iron-pnictide superconductor. (c) Electronic DOS. Reprinted with permission from [28].

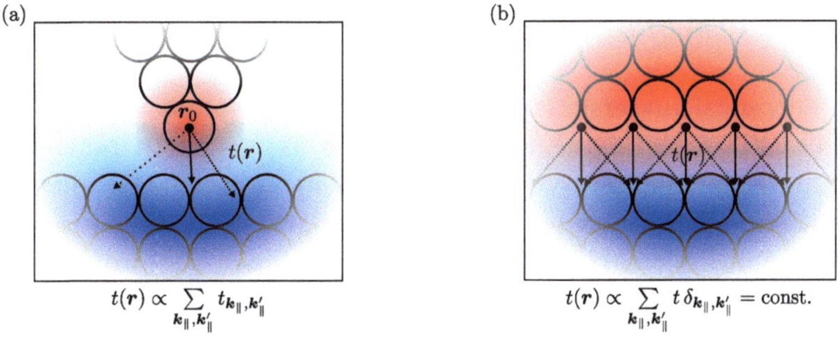

Figure 2.9: Tunnelling in STM and Planar Junctions: The difference in tunnelling processes due to the tunnelling geometry are schematically illustrated for an STM geometry of tip and sample (a) and a planar junction (b). In both, the occupied (initial) local density of states (LDOS) of the top electrode and the unoccupied (final) local density of states (LDOS) of the bottom electrode are shaded in red and blue respectively. The tunnelling amplitude $t(\boldsymbol{r})$ is largest where the electronic eigenfunctions of one lead are most strongly disturbed by the potential of the opposite lead (for the STM geometry this is the point $\boldsymbol{r}_0$).

keep in mind that the spin-fermion model fails to describe nodal parts in cuprate superconductors correctly. In addition to that, a full spin-gap is not expected for them, because of the coupling between spin-fluctuations and non-gapped electrons.

2.6 Quantum Tunnelling in NIS Junctions

The BCS theory was a major success in explaining the macroscopic characteristics of superconductors, i.e. the observed jump in the specific heat, the dissipationless current and the Meißner and isotope effect. Another confirmation of one of the theory's corner stones followed three years later, when Giaver [19] realized that the slope $\mathrm{d}I/\mathrm{d}U$ of the tunnelling current I in normal metal-insulator-superconductor (NIS) sandwich structures under an applied voltage U resembled the BCS DOS for quasiparticle (QP) excitations. Not only did this experiment cement the existence of the superconducting gap in the DOS of the superconductor, it also paved the way for tunnelling experiments to become one of the primary techniques for investigating the low energy DOS in not only superconductors, but conducting solids in general. Therefore, historically, the exploration of the energy spectrum of superconductors and development of tunnelling theory are closely intertwined.

2.6.1 Theoretical Framework of the Tunnelling Geometry

A good theoretical understanding of Giaver's observation was developed by Bardeen [99]. Describing the insulator region as a smooth potential barrier (Wentzel–Kramers–Brillouin (WKB) approximation), he could show, that an overlap of the many-particle wavefunctions from the normal metal and the superconductor leads to tunnelling normal electrons with energy eU at a rate proportional to the superconductor's QP DOS at $E_F + eU$. In order to keep this section concise, only formulas appropriate for the interpretation of experimental data obtained in a STM will be presented. Wavefunction and derived functions of the atomically sharp tip electrode are denoted with the index t, for the sample they carry the index s.

In contrast to planar junctions, the potential barrier separating the two leads in an STM is the vacuum. This simplifies things because the vacuum is perfectly free of states that could interfere with the tunnelling electrons. The height of the potential barrier linearly changes from the work function Φ_t of the tip to Φ_s of the sample. The quantum mechanical tunnelling intensity $|t|^2$ and therefore also the tunnelling current

$$I \propto |t|^2 \propto \exp(-2\kappa d) \tag{2.63}$$

is in a WKB approximation exponentially dependent on the tip-sample distance d with $\kappa = \sqrt{2m(\bar{\Phi} + eU)}/\hbar$, where $\bar{\Phi} = (\Phi_s + \Phi_t)/2$ is the averaged barrier height. An exponential dependence of the tunnelling current on the applied bias voltage only arises when the bias voltage is comparable to the work function, typically several electronvolt. For small bias voltages, $eU \ll \bar{\Phi}$, this effect is negligible and κ can be considered bias independent.

The atomic sharpness of the tip electrode has two important consequences: First, Tersoff and Hamann [100] could show that for small bias voltages and a metallic tip with only s-wave states, the current through the tip-sample tunnel junction is given by

$$I \propto \int_0^{eU} \nu_\mathrm{s}(\boldsymbol{r}_0, E_F + \epsilon)\mathrm{d}\epsilon. \tag{2.64}$$

The tunnelling current is, hence, directly proportional to the integrated local density of states (LDOS) $\nu_\mathrm{s}(\boldsymbol{r}, E)$ of the sample in the energy interval $[E_F, E_F + eU]$ at the tip position $\boldsymbol{r}_0$. This is schematically illustrated in the STM geometry shown in Fig. 2.9(a). The local density of occupied (initial) states of the tip electrode are shaded in red and the local density of unoccupied (final) states of the sample are shaded in blue. The total overlap is mostly determined by the overlap at $\boldsymbol{r}_0$ due to the spherical wave nature of the tip state.

Second, the contribution of inelastic tunnelling events to the total current is an order of magnitude higher than in planar tunnel junctions [73]. That is because the tip breaks the lateral translation symmetry which usually enforces the conservation of in-plane momentum of the tunnelling electron [101] in planar junctions. By removal of this condition, the phase space for a bosonic excitation grows significantly [70]. The different tunnelling paths in real space are shown next to each other in Fig. 2.9 for the STM geometry (a) and the planar junction geometry (b). In the STM geometry, tunnelling happens in a conical region below the tip, although the tunnelling amplitude $t(\boldsymbol{r})$ quickly falls off for less vertical tunnelling paths due to the smaller integral overlap of the wavefunctions of tip and sample. In a perfect planar junction, the tunnelling paths are constrained to lead-to-lead transitions that fulfill $\boldsymbol{r}_{\|} \to \boldsymbol{r}_{\|} + \boldsymbol{G}_{\|}$, where $\boldsymbol{G}_{\|}$ is a in-plane lattice vector, due to the conservation of in-plane momentum. The tunnelling amplitude $t(\boldsymbol{r})$ is roughly constant for all points along the lead edges and mainly determined by vertical tunnelling paths [101].

From Eq. (2.63) and (2.64) and Fig. 2.9(a) we get a good idea of how topographic images of the sample surface are reconstructed in the constant-current mode of an STM (see also Sec. 3.1). In order to gain a better understanding of what kind of information is contained in a bias spectroscopy of the tunnelling current, i.e. a $I(U)$ curve, let us neglect the explicit energy dependence of the tunnelling amplitude t and drop the local dependence for now. We can keep them in the back of our minds for later. First, let us only take a look at elastic tunnelling events.

2.6.2 Elastic Tunnelling

First, let us introduce the dimensionless electronic DOS

$$\tilde{\nu}_{\mathrm{s/t}}(\epsilon) = \frac{\nu_{\mathrm{s/t}}(\epsilon)}{\nu_F^{\mathrm{s/t}}}, \tag{2.65}$$

where ν_F denotes the DOS at the Fermi energy. The elastic tunnelling current is given by [70, 102]

$$I^{\mathrm{el}}(U) = \frac{\sigma_0}{e} \int_{-\infty}^{\infty} \mathrm{d}\epsilon \, \tilde{\nu}_{\mathrm{s}}(\epsilon)\tilde{\nu}_{\mathrm{t}}(\epsilon - eU) \left[S(\epsilon - eU, \epsilon) - S(\epsilon, \epsilon - eU)\right], \tag{2.66}$$

$$S(\epsilon_1, \epsilon_2) = n_F(\epsilon_1)\left(1 - n_F(\epsilon_2)\right). \tag{2.67}$$

The conductance constant $\sigma_0 = 4\pi e^2 \nu_F^{\mathrm{t}} \nu_F^{\mathrm{s}} |t^{\mathrm{el}}|^2$ is the purely elastic conductance for a constant sample and tip DOS and $n_F(\epsilon) = [\exp((\epsilon - \mu)/k_B T) + 1]^{-1}$ are the Fermi-Dirac distributions of states with energy ϵ. Additionally, we assumed the tip DOS to be constant and the elastic tunnelling matrix element $|t^{\mathrm{el}}|^2 = |\langle \psi_{\mathrm{t}}(\boldsymbol{k}', \epsilon)|\hat{T}|\psi_{\mathrm{s}}(\boldsymbol{k}, \epsilon)\rangle|^2$

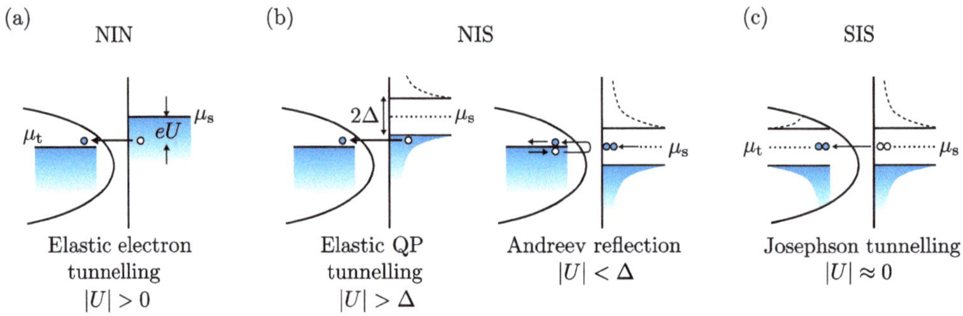

Figure 2.10: Elastic Tunnelling Processes: Schematic of elastic tunnelling processes between tip (left) and sample (right), where blue spheres depict electrons and white spheres holes or missing electrons. (a) Elastic electron tunnelling in an NIN junction. (b) Elastic QP tunnelling and Andreev reflection in an NIS junction. (c) Josephson tunnelling in an SIS junction.

to be energy and momentum independent. The differential conductance is then given by

$$\sigma^{\text{el}}(U) = \frac{\mathrm{d}I^{\text{el}}(U)}{\mathrm{d}U} = -\sigma_0 \int_{-\infty}^{\infty} \mathrm{d}\epsilon \tilde{\nu}_{\text{s}}(\epsilon) n_F'(\epsilon - eU). \tag{2.68}$$

For $T \to 0$, it follows $n_F'(\epsilon - eU) \to -\delta(\epsilon - eU)$. Therefore, for very low temperatures, the elastic conductance is directly proportional to the density of states of the sample $\nu_{\text{s}}(eU)$ at the energy eU. Else, it is described by the convolution integral in Eq. (2.68), i.e. the DOS is "smeared out" on the scale $k_B T$. Ramping the bias voltage U and recording the differential conductance $\sigma = \mathrm{d}I/\mathrm{d}U$, essentially mirrors the sample DOS around the Fermi energy.

The simple elastic tunnelling process of normal electrons between a metallic tip and sample through the vacuum barrier is schematically shown in Fig. 2.10(a). If the sample is superconducting (Fig. 2.10(b)), QPs can only tunnel directly into or out of the sample for $e|U| > \Delta$, because of the superconducting gap in the sample's QP DOS. Only for a small tunnelling barrier, so when going continuously from the tunnelling to the point-contact regime, Andreev reflections yield finite current for $e|U| < \Delta$. During an Andreev reflection, a hole (electron) in the tip is retro-reflected as an electron (hole) with opposite momentum and spin at the boundary to the superconductor, which leads to the annihilation (creation) of a Cooper pair in the superconducting sample. If tip and sample are superconducting (Fig. 2.10(c)), then Cooper pairs can tunnel between them through the vacuum barrier, leading to sharp spike in the tunnelling current for $U = 0$.

2.6.3 Inelastic Tunnelling

Although mainly consisting of elastic contributions, the total conductance also contains inelastic contributions, i.e. $\sigma = \sigma^{\mathrm{el}} + \sigma^{\mathrm{inel}}$. In STM, their weight is not insignificant. As inelastic tunnelling spectroscopy (IETS) on Pb films shows, the inelastic contributions can make up 9 % of the total conductance [73, 74]. The inelastic tunnelling current is given by [70]

$$I^{\mathrm{inel}}(U) = \frac{\sigma_0}{e} \frac{1}{\nu_{\mathrm{s}}^F D^2} \int_{-\infty}^{\infty} \mathrm{d}\epsilon \int_0^{\infty} \mathrm{d}\omega \times \tag{2.69}$$

$$\big(\alpha^2 F(\omega)\tilde{\nu}_{\mathrm{s}}(\epsilon)\tilde{\nu}_{\mathrm{t}}(\epsilon-\omega-eU)S(\epsilon-\omega-eU,\epsilon)n_B(\omega) \tag{2.69a}$$

$$-\,\alpha^2 F(\omega)\tilde{\nu}_{\mathrm{s}}(\epsilon)\tilde{\nu}_{\mathrm{t}}(\epsilon-\omega-eU)S(\epsilon,\epsilon-\omega-eU)[1+n_B(\omega)] \tag{2.69b}$$

$$+\,\alpha^2 F(\omega)\tilde{\nu}_{\mathrm{s}}(\epsilon)\tilde{\nu}_{\mathrm{t}}(\epsilon+\omega-eU)S(\epsilon+\omega-eU,\epsilon)[1+n_B(\omega)] \tag{2.69c}$$

$$-\alpha^2 F(\omega)\tilde{\nu}_{\mathrm{s}}(\epsilon)\tilde{\nu}_{\mathrm{t}}(\epsilon+\omega-eU)S(\epsilon,\epsilon+\omega-eU)n_B(\omega)\big)\,. \tag{2.69d}$$

ω is the energy of the absorbed/emitted boson with occupation number $n_B(\omega)$, the Bose-Einstein distribution function. D is a characteristic cut-off energy of the boson spectrum, or in other words, the band width of off-shell states. α is the electron-boson coupling constant and $F(\omega)$ the boson spectral density. For exclusively phononic excitations, $\alpha^2 F(\omega)$ is directly proportional to the Eliashberg function and will often be used synonymously, although slight differences exist (for details see Ref. [70]).

Eq. (2.69) comprises the general inelastic tunnelling processes involving one boson of energy ω. The processes are:

(2.69a) A tip electron tunnels into the sample. A boson in the sample is absorbed in the process.

(2.69b) A sample electron tunnels into the tip. A boson in the sample is emitted in the process.

(2.69c) A tip electron tunnels into the sample. A boson in the sample is emitted in the process.

(2.69d) A sample electron tunnels into the tip. A boson in the sample is absorbed in the process.

Processes (2.69a) and (2.69d) are essentially absent at $k_B T \ll \omega$ because the thermal equilibrium occupation of these bosonic excitations is zero. For $k_B T \ll eU$ only one

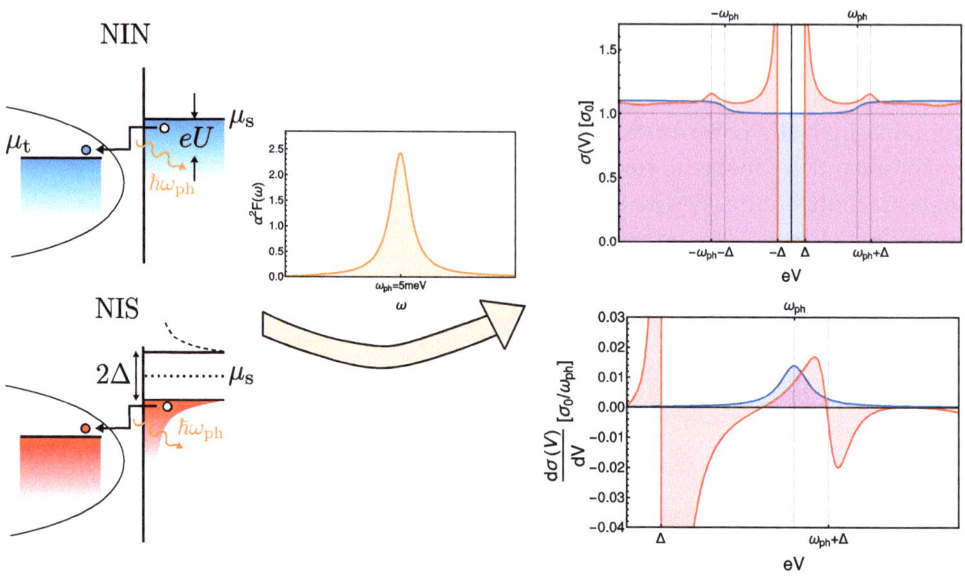

Figure 2.11: Signatures of Inelastic Tunnelling: On the left side, the inelastic tunnelling process of electrons (quasiparticles) in an NIN (NIS) junction are schematically depicted like in Fig. 2.10(a,b). The excitation of a single phonon mode with spectral density $\alpha^2 F(\omega)$ (orange curve) centred at ω_{ph} leads to the total tunnelling conductance $\sigma = \sigma^{\text{el}} + \sigma^{\text{inel}}$ and its derivative on the right. The curves for the NIN/NIS junction are shown in blue/red. Partly taken from Ref. [70]

of the remaining processes, (2.69b) or (2.69c), is present, depending on the direction of the bias voltage. For positive voltage the inelastic conductance is then given by

$$\sigma^{\text{inel}}(U > 0) = \frac{\mathrm{d}I^{\text{inel}}(U > 0)}{\mathrm{d}U} \tag{2.70}$$

$$\approx -\frac{\sigma_0}{\nu_{\text{s}}^F D^2} \int_0^\infty \mathrm{d}\omega \alpha^2 F(\omega)[1 + n_B(\omega)] \int_{-\infty}^{\infty} \mathrm{d}\epsilon \tilde{\nu}_{\text{s}}(\epsilon) \frac{\partial S(\epsilon, \epsilon + \omega - eU)}{\partial(-eU)} \tag{2.71}$$

$$\overset{T \to 0}{\to} \frac{\sigma_0}{\nu_{\text{s}}^F D^2} \int_0^{eU} \mathrm{d}\omega \alpha^2 F(\omega) \tilde{\nu}_{\text{s}}(eU - \omega). \tag{2.72}$$

After evaluating the Bose-Einstein and Fermi-Dirac distributions for zero temperature, the inelastic conductivity is given by a convolution integral of the boson spectral density and the sample DOS over the real-valued boson frequencies $\omega \in [0, eU]$. In the normal conducting state ($\tilde{\nu}_{\text{s}} = 1$), the second derivative of the tunnelling current is then proportional to $\alpha^2 F(\omega)$. In the superconducting state, the non-constant sample DOS has to be respected.

In order to develop an intuitive understanding of Eq. (2.68) and (2.72), calculated conductance signatures for normal metal-insulator-normal metal (NIN) and NIS tunnelling, involving a single Gaussian-shaped phonon mode at $\omega_{\rm ph}$, are shown in Fig. 2.11. Elastic plus inelastic tunnelling via emission of a phonon of frequency $\omega_{\rm ph}$, with the spectral density $\alpha^2F(\omega)$ shown in orange, in the normal/superconducting sample lead to the differential conductance curve shown in blue/red. For the normal conducting sample, the differential conductance σ exhibits a step at $\omega_{\rm ph}$ and a peak at the same energy in $\mathrm{d}\sigma/\mathrm{d}U$ due to the inelastic tunnelling contribution. For the superconducting sample, σ shows a small (gaussian broadened) copy of the coherence peak at $\omega_{\rm ph} + \Delta$ and a change of sign at the same energy in $\mathrm{d}\sigma/\mathrm{d}U$. Again, this is due to the inelastic tunnelling contribution. The increase in differential conductance for $\omega_{\rm ph} < eU < \omega_{\rm ph} + \Delta$ and the oscillations at higher energies are due to strong electron-phonon coupling effects, which alter the DOS, so the elastic conductance, of the superconductor.

This simple example already goes to show that in the presence of strong electron-boson coupling and significant inelastic tunnelling into the superconductor, a determination of the Eliashberg function from just the tunnelling spectrum becomes difficult, because $\alpha^2F(\omega)$ enters the elastic and inelastic part. For large inelastic contributions in scanning tunnelling spectroscopy (STS) measurements on conventional superconductors, the Eliashberg function can simply be determined through $\mathrm{d}^2I/\mathrm{d}U^2$ in the normal state. For negligible inelastic contributions like in planar junction experiments, it can still be determined through McMillan and Rowell's inversion procedure [72], which fits Eliashberg theory parameters iteratively until the DOS matches the tunnelling data. For unconventional superconductors, both of the above-mentioned methods fail. For one, the spectral density of the bosonic glue undergoes drastic changes from the normal to the superconducting state and for a second, the electron-boson coupling is not of phononic nature and the use of Migdal's theorem, which the Migdal-Eliashberg theory builds up on, is not justified. In Chap. 4, we will see, how scanning tunnelling spectra on unconventional superconductors can be used to qualitatively reconstruct the boson spectral density when the contribution of inelastic contributions outweighs the fine structure of the elastic conductance.

3 Experimental Setup and Methods

Scanning tunnelling microscopy is a surface science technique that grants access to the local density of electronic states at a conducting sample surface. It is therefore the ideal choice to study electronic properties in real space with local information down to the atomic scale. The resolution of this technique is highly dependent on its technical realization, e.g. vibration insulation, signal amplification, and measurement temperature. For the experiments in this work, a low temperature and ultra-clean sample surface were in all cases a prerequisite. This chapter first describes working principle and operational procedures of the experimental setup, then introduces the setups of the two used scanning tunnelling microscopes including the ultra-high vacuum (UHV) system, the cryogenic system and the decoupling from various noise sources, and lastly presents sample and probe preparation.

3.1 Scanning Tunnelling Microscopy and Spectroscopy

The first prototype of a scanning tunnelling device and precursor for the scanning tunnelling microscope, we know today, was the topografiner, developed by R. D. Young and co-workers between 1965 and 1971 [103]. A metallic stylus, moveable by three-axes piezo actuators, scanned the sample surface at a constant distance recreating the surface topography. However, due to the lack of sufficient vibration

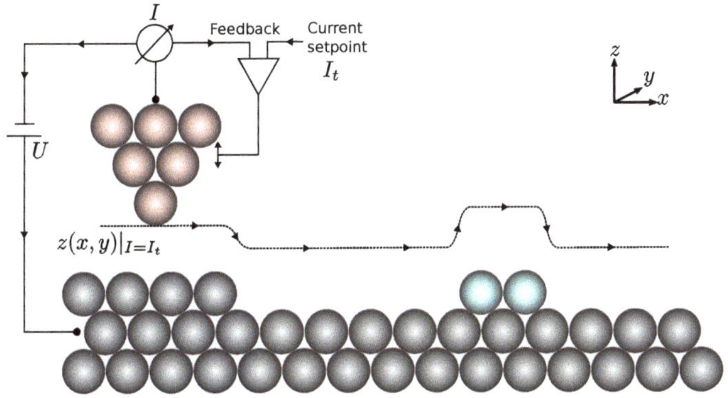

Figure 3.1: Constant Current Mode: In the constant current mode of the STM, a feedback-loop of the tunnelling current regulates the tip-sample distance. The topography $z(x,y)|_{I=I_t}$ contains not only a height profile but also chemical contrast due to its sensitivity to the electron density at the surface.

insulation the tip-sample distance could not be kept small enough to be in tunnelling range but the instrument was operated in field emission range. In 1981/1982 Binnig and Rohrer presented the first version of the scanning tunnelling microscope (STM) [104], a technical improvement of the topografiner, that bettered other real space imaging techniques at the time, like field ion microscopy (FIM) and scanning electron microscopy (SEM), in terms of vertical resolution while providing similar lateral resolution. In 1986, Binnig and Rohrer received the Nobel Prize in physics for their invention - a testament to how important the delicate insulation of this sensitive instrument is.

Working principle In an STM, an atomically sharp metal tip is held several angstroms above the sample surface, close enough for electrons to tunnel through the vacuum barrier between sample and tip. The tunnelling current, usually pA to several nA is driven by a bias voltage, which is usually much smaller than the sample and tip work function. Due to the exponential dependence between tunnelling current and tip-sample distance (WKB-approximation), the vertical resolution of the STM can be in the pm range. These small distance adjustments are realized by a piezo scanner tube. With its four-segment electrode contact, a fine motion in all three directions is possible with good linearity and its dimensions ensure a high resonance frequency. For a coarse alignment in horizontal (x) and vertical (z) direction, the sample stage and the scanner respectively are moved by stack piezo actuators that perform slip-stick motion [105].

Electronics In the STM designs used in this work, the bias voltage U is applied to the sample and the tip is connected to a transimpedance amplifier that converts the tunnelling current I between tip and machine ground to FPGA readable voltages. Typical amplifications range from 10^7 V/A to 10^9 V/A. The FPGA card is the heart of the real-time controller (RC) by Nanonis. It includes an analog-to-digital converter (ADC) and digital-to-analog converter (DAC) making an easy real-time control of the STM tip possible via a LabView based computer interface. In the constant-current mode, the tip-sample distance is controlled via a feed-back loop: The user can set a set-point current I_t, e.g. $1\,\mathrm{nA}$, and the voltages on the piezo scanner are then, in real-time, adjusted continuously such that the tunnelling current always matches the set-point current. This is schematically demonstrated in Fig. 3.1. In the constant-height mode, the tip-sample distance is kept constant (feed-back loop off) at z_t.

Topographic scan image After piezo voltages are calibrated to a tip displacement, a rectangular area on the sample surface can be scanned in constant-current-mode by applying sawtooth voltage signals to the leads of the piezo scanner responsible for x and y fine motion. A map of the value $z(x,y)|_{I=I_t}$ at fixed bias voltage is what is called the topographic scan image. A map of the tunnelling current $I(x,y)|_{z=z_t}$ on the other hand is obtained in constant-height mode. Usually, a topographic scan first ensures a flat surface on which a scan in constant-height mode can then be performed safely.

Bias spectroscopy Bias spectroscopy is the recording of an $I(U)$ curve at a specific tip position (x,y,z). During the voltage ramp, the feed-back loop is turned off. A (line) grid spectroscopy consists of bias spectroscopies at several points defined on a (1D) 2D grid on the surface. In scanning tunnelling spectroscopy (STS), one is often interested in the first derivative of the tunnelling current with respect to bias voltage, $\mathrm{d}I/\mathrm{d}U$. Since the tunnelling current carries noise of all frequencies (especially low frequencies can be a nuisance), the use of a lock-in amplifier to filter out most noise frequencies for the cost of a little bit of energy resolution is usually favoured over a numerical derivative.

Lock-in amplifier The goal of this lock-in technique is to extract the first derivative of the tunnelling current with respect to the bias voltage as noise-free as possible. First, the DC tunnelling bias U_0 is modulated by addition of an AC voltage $U_{\mathrm{mod}}\cos(\omega t)$. As a result, the tunnelling current also picks up an AC component at the frequency ω which is proportional to the first derivative $\mathrm{d}I/\mathrm{d}U|_{U_0}$ and the amplitude of the modulation U_{mod}. In a second step, the tunnelling current signal is demodulated in the lock-in amplifier by multiplication with a reference signal of the frequency ω and time-averaging. This averaging cancels out the signal and noise at all frequencies that are different from ω. Only the signal and noise at ω remain. The phase of the tunnelling current and the reference signal can be aligned as follows: Out of tunnelling contact, the main contribution to the AC current at ω is due to the capacitive impedance X_c of the junction. Aligning the reference phase to maximize this contribution and rotating it by another $90°$ places it in line with the AC current due to ohmic resistors. The by far largest ohmic resistance is the tunnelling resistance. Hence, with this phase alignment the demodulated signal is maximally sensitive to the tunnelling current and blind to capacitive and inductive contributions.

Energy resolution In order to determine the maxmimum energy resolution of the Dilution-STM at $36\,\mathrm{mK}$, the $\mathrm{d}I/\mathrm{d}U$ feature of a DC Josephson current, or alternatively the zeroth order Andreev bound state, was measured. Even though this feature has a functional shape governed by the dissipation of Cooper pairs through inelastic

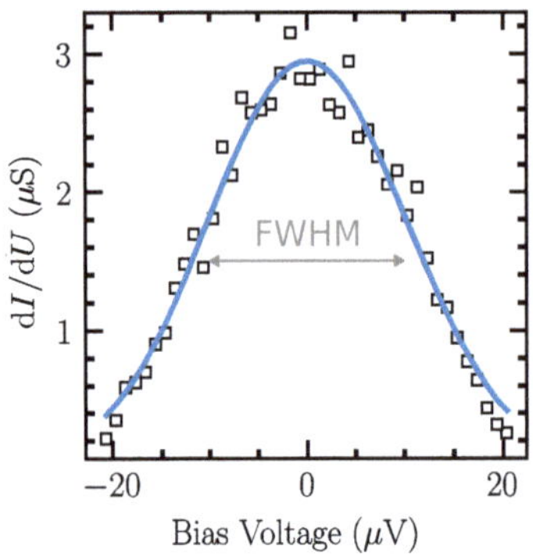

Figure 3.2: Energy Resolution Limit: Differential conductance feature of a DC Josephson current (black squares). The spectrum was measured in a Pb-Pb tunnelling contact at $T = 36\,\mathrm{mK}$ and $U_{\mathrm{mod}}^{\mathrm{PK}} = 5\,\mu\mathrm{V}$. A normal distribution was fitted to the data points (blue line).

processes including thermal photons [106], we can approximate it by a delta function centred at zero voltage[1]. Since the Cooper pair tunnelling signature is not broadened by a Fermi distribution in the initial and final states, an analysis of this feature lets us decouple pure distribution function effects from thermal voltage fluctuations by the junction capacitance and bias voltage modulation. We prepared a Pb tip by soft dipping a tungsten tip into a Pb(111) single crystal and measured the zero energy peak in $\mathrm{d}I/\mathrm{d}U$ on the same Pb(111) crystal at small tip-sample distance. The used peak modulation voltage was $U_{\mathrm{mod}}^{\mathrm{PK}} = 5\,\mu\mathrm{V}$. The bias spectroscopy is shown in Fig. 3.2. The blue line shows the fitted function of Gaussian shape with a full width at half maximum (FWHM) of

$$\mathrm{FWHM} = \sqrt{(2U_{\mathrm{mod}}^{\mathrm{RMS}})^2 + 2\ln(2)(2\sigma_U^{\mathrm{el}})^2}. \tag{3.1}$$

From the fit it was found that $\sigma_U^{\mathrm{el}} = (9.4 \pm 0.2)\,\mu\mathrm{V}$. $U_{\mathrm{mod}}^{\mathrm{RMS}}$ is the root mean square (RMS) value of the AC modulation related to the peak (PK) value by $U_{\mathrm{mod}}^{\mathrm{RMS}} = U_{\mathrm{mod}}^{\mathrm{PK}}/\sqrt{2}$. The temperature broadening of $\mathrm{d}I/\mathrm{d}U$ features can (with minor errors) also be approximated by a Gaussian distribution. The full energy resolution (FWHM) of the instrument at milli-kelvin temperatures is finally composed of the three constituents in the following way

$$\Delta\varepsilon = \sqrt{(3.2k_{\mathrm{B}}T)^2 + (2eU_{\mathrm{mod}}^{\mathrm{RMS}})^2 + ((22.1 \pm 0.5)\,\mu\mathrm{eV})^2}. \tag{3.2}$$

The energy resolution of the instrument below $T \sim 80\,\mathrm{mK}$ and a modulation voltage of $U_{\mathrm{mod}}^{\mathrm{PK}} \sim 15\,\mu\mathrm{V}$ is mainly limited by what is thought to be voltage fluctuations due to thermal charge fluctuations in opposite sides of the tunnelling junction leads.

[1] The spectral function $P_C(U)$ has a FWHM $\leq 2\,\mu\mathrm{V}$, which is significantly lower than the limitation of energy resolution due to thermal voltage noise.

Differential conductance map A topographic scan and a map of differential conductance $\mathrm{d}I/\mathrm{d}U|_{I=I_t}$ at bias voltage U can be obtained simultaneously when the lock-in amplifier is running during the scan and the integration time per pixel is long enough. Thus, differential conductance maps are usually recorded at lower tip speeds.

Multi-pass image Of special importance in this work are so-called multi-pass images. Maps of differential conductance at energies eU smaller than the superconducting gap Δ are not obtainable with a turned-on feedback-loop. In that case the tip would go into direct contact with the sample as soon as the sample exhibits a full superconducting gap because of a vanishing tunnelling current. Full grid spectroscopies are time-consuming and if one is only interested in a small number of energies but a fine spatial resolution, a multi-pass image is the best choice. First, the tip scans a row of pixels forward and backward at a bias voltage larger than Δ/e and records the profile $z(x,y)_{I=I_t}$ along this line. Then, the tip plays back the previously recorded profile at an energy $eU < \Delta$ and records the lock-in signal. This is also advantageous compared to the constant-height mode because areas with larger height distribution are not a problem. In order for this method to work well, high mechanical stability, low thermal drift and small piezo creep are important. Vertical tip offsets in the play-back phase can either decrease the tip-sample distance ($z_{\text{off}} < 0$) for an enhanced lock-in signal or increase it ($z_{\text{off}} > 0$) for an additional safety buffer but smaller lock-in signal.

3.2 Low Temperature Scanning Tunnelling Microscope

In this work, two different setups were used: the *dilution scanning tunnelling microscope (D-STM)* and the *Joule-Thomson scanning tunnelling microscope (JT-STM)* who owe their name to their central cooling system. Both systems including UHV setup, cryostat, magnetic coil and scanning tunnelling microscope are home-built after original designs in the group of Prof. Wulf Wulfhekel [107, 108]. As the D-STM was the primary setup used for all experiments except the one described in Chap. 4, and is the more intricate one of the two, it will serve as an example to explain the UHV setup, cryostat and STM design. Differences to the JT-STM are highlighted in Sec. 3.2.2.

3.2.1 Dilution Scanning Tunnelling Microscope

UHV setup The vacuum setup follows a classical three chamber design: The three chambers - preparation chamber, STM chamber and vacuum load-lock - are placed in an L-shape in the horizontal plane. The preparation and STM chamber are connected via a gate valve and placed in line with a shared coolable, horizontal manipulator that is mounted in the preparation chamber. The vacuum load-lock chamber is mounted to the preparation chamber at an angle of 90° with respect to the STM chamber connection. The load-lock has a lid, that can be opened once this chamber is vented and a small vertical manipulator in the load-lock can pass samples on into the preparation chamber. Additionally, the preparation and STM chamber possess small wobble-stick manipulators for more intricate manipulation of samples in UHV.

All chambers have a connection to the *barrel*, a cylindrical container that is pumped by a roughing pump. The barrel acts as a rough pre-vacuum for each turbomolecular pump (TMP). The load-lock, preparation and STM chamber can all be pumped with separate TMPs. The TMPs are all turned off during an STM measurement due to their vibrational noise. Instead, the preparation and STM chamber are constantly pumped by ion getter pumps which work based on the principle of physisorption and chemisorption and are thus free of mechanical noise. On top of that, the preparation chamber is equipped with a titanium sublimation pump (TSP) and the STM chamber contains a non-evaporable getter (NEG) pump, a porous reactive material with a very high surface area, ideal for the adsorption of hydrogen. Like this, base pressures below 10^{-10} mbar can be reached in the preparation and STM chamber. It should be noted that the cryostat also acts as a large cryopump. The pressures are therefore mostly in the lower 10^{-11} mbar range inside the STM chamber and even another two orders of magnitude lower inside the closed 1 K shield where the STM sits. The chamber pressures are constantly monitored by ionization gauges and ion currents in the getter pumps.

Peripheral equipment mounted on the preparation chamber includes a sputter gun (differentially pumped via load-lock), a combined instrument for low energy electron diffraction (LEED) and Auger electron spectroscopy (AES), five electron-beam evaporators for molecular beam epitaxy (MBE) and a quadrupole mass spectrometer (QMS).

Cryostat The wet cryostat shown in Fig. 3.3(a) features a 13 L liquid helium tank below a 23 L liquid nitrogen tank. Heat transfer by conduction is kept low by thermal insulation at their mounting point on the top flange while heat transfer by radiation is kept low by vacuum surrounding and additional radiation shields. Since radiational heat transfer from a larger temperature to a lower temperature

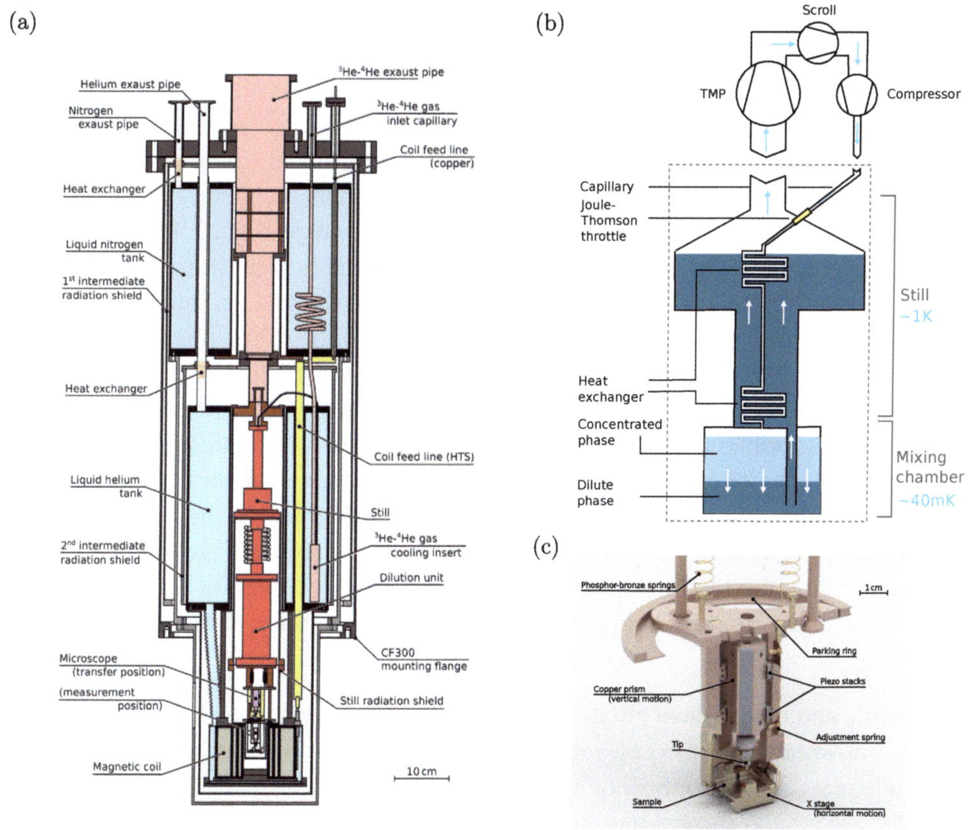

Figure 3.3: D-STM cryostat and STM: (a) Cross-sectional view of the D-STM cryostat including the STM and magnetic coil. (b) Schematic depiction of the ^{3}He/^{4}He condensation, phase separation and gas circulation. (c) 3D rendered image (cut) of the STM body. (a,c) have been reprinted from [108] with permission of AIP Publishing.

reservoir is proportional to $(T_>^4 - T_<^4)$ it is optimal to use several radiation shields which are equally distributed in temperature than one shield where the temperature gradient is large. This cryostat design includes five concentric radiation shields: an intermediate radiation shield (180 K), cooled by evaporating liquid nitrogen, a shield physically connected to the liquid nitrogen tank (77 K), a second intermediate radiation shield (28 K), cooled by evaporating liquid helium, a shield physically connected to the liquid helium tank (4.2 K) and a still radiation shield (1 K).

The base temperature of 25 mK in the microscope is reached with the built-in commercial dilution unit by Bluefors Cryogenics. The coolant in such a dilution refrigerator is a ^{3}He/^{4}He mixture. Below a critical temperature of $T_c = 870$ mK (for a ^{3}He concentration of $x_c = 0.67$) the mixture separates into two phases - a dilute and a concentrated phase [109]. The concentrated phase is almost pure ^{3}He, the dilute phase contains a low amount of ^{3}He and the content cannot fall below $x_0 = 6.5\,\%$. A dilution refrigerator makes use of this constraint to the phase equilibrium by removing ^{3}He from the dilute phase. In order to satisfy the concentration constraint in the dilute phase, ^{3}He atoms from the concentrated phase are forced to cross the phase boundary in an endothermic process [109]. A schematic picture of the dilution unit in the steady state is shown in Fig. 3.3(b). The flow direction of ^{3}He is shown as light blue arrows in the gas and as white arrows in the liquid phase. In the steady state almost all of the gas mixture is condensed inside the still and mixing chamber. The concentrated phase is lighter and thus floats atop the dilute phase from which ^{3}He is constantly removed by a TMP. Much less ^{4}He than ^{3}He is removed from the dilute phase because of its significantly lower vapor pressure. The removed ^{3}He is constantly resupplied in a closed gas cycle. In order to reach the steady state, the gas mixture has to be readily precooled. It is pressed through a small capillary with the help of a compressor and then precooled by heat exchangers in the liquid nitrogen tank and liquid helium tank (see Fig. 3.3(a)). Inside the still it is again precooled by the evaporating mixture and experiences an isoenthalpic Joule-Thomson expansion which brings the mixture below T_c when entering the mixing chamber.

STM body and mechanical noise reduction Most parts of the STM body, shown in Fig. 3.3(c), are made from copper or gold-plated copper for quick thermalization. The only real thermal connection is established by pure silver wires to the mixer. The microscope temperature is monitored by a calibrated RuO_2 thermometer, mounted on the STM body. $15 \times 15\,\mathrm{mm}$ sample plates place the sample directly under the tip. The optical access to the junction in the parked position of the microscope (elevated position above coil) allows for a quick in-situ exchange of sample or tip.

In the unparked position, the microscope hangs on three soft springs with sub-Hz resonance frequency to enhance vibrational damping in the high frequency range. For insulation from low frequency mechanical noise, the whole machine, including all chambers and cryostat, sits on pneumatic vibration isolators by Newport. Vibrational noise damping from the pumps of the ^{3}He circulation is realized by decoupling bellows and Helmholtz resonators mounted on the pump line.

Magnetic coil and heater Of special importance in this work is the influence of magnetic fields and temperature on superconductivity. The D-STM features a superconducting NbTi-based coil, suspended in liquid helium, with a maximum field of $6.5\,\mathrm{T}$. In the unparked position the STM hangs in the axial centre of the superconducting magnet giving access to the maximum magnetic field in vertical (z) direction. The electrical power to the coil is supplied through shielded copper wire up to the liquid nitrogen tank at 78 K and through high temperature superconducting wires made from rare earth barium copper oxide (REBCO) further down in the cryostat.

The measurement temperature can be increased by using a 10 kΩ resisitive heater mounted on the STM body.

3.2.2 Joule-Thomson Scanning Tunnelling Microscope

The JT-STM's cryostat lacks a dilution refrigerator and instead uses the Joule-Thomson effect to liquify pure ^{3}He which, in contrast to pure ^{4}He, is not in a suprafluid phase at this temperature. The lowest achievable temperature at the time of the experiments was 630 mK. In the parked position, the STM has a direct thermal connection to the liquid helium bath enabling a quick cooldown. In the unparked position the STM hangs in the axial centre of a 3 T split-coil magnet. The STM head itself has an upside down design compared to the one in the D-STM: The z prism and scanner approach from the bottom and the samples are facing down.

3.3 Tip and Sample Preparation

3.3.1 Tip Preparation

Electrochemical Etching The ca. 3 mm long tips were electrochemically etched from a 150 µm thick tungsten wire using a version of the DC drop-off technique [110]. The setup is shown in Fig. 3.4(a). The tungsten wire is the anode and the gold ring the cathode. An alkaline solution of NaOH ($c \sim 1\,\mathrm{mol/L}$) is applied to form a meniscus inside the gold ring. Upon exceeding the activation voltage of 1.43 V, soluble tungstate ions are formed at the anode in the following electrochemical reaction [110]:

$$\mathrm{W(s)} + 2\mathrm{OH}^- + 2\mathrm{H_2O} \rightarrow \mathrm{WO_4^{2-}} + 3\mathrm{H_2(g)}. \quad (3.3)$$

In reality, the required voltage for the oxidation to start is slightly higher than calculated from standard potentials. In this work, a voltage between 1.7 V and 2.0 V was frequently used. An ammeter is put in series to observe the ion current. As soon as it drops too low, the solution is reapplied to the meniscus to restore the original concentration of OH^- ions. Through the dissolution of tungstate ions, the wire thins at the electrolyte/tungsten interface until it is so thin that the lower end is pulled out by gravitation. The falling part of the tungsten wire is caught by a cushion of shaving foam and used as the tip since the reaction is immediately stopped for this part after the severing. It has been reported that the upper part of the wire is blunted by the continued etching in the meniscus after separation from the lower part [111].

Flashing After electrochemical etching, the tips were glued inside a tip-holder by silver glue and transferred into the UHV chamber. In order to remove its $5 - 10\,\mathrm{nm}$ thick, insulating WO_3 layer [112], it is locally heated by electron bombardment in several short flashes. A positive voltage of $1\,\mathrm{kV}$ on the tip accelerates emitted electrons from a filament onto it. The maximum emission current between filament and tip is limited to $100\,\mathrm{mA}$. The voltage at the filament is turned up quickly until the maximum heating power is reached and then immediately turned off. The maximum heating power is only present for few milliseconds. These short bombardment periods ensure a very local heating at the tip by the concentrated electric field. At the heated tip end WO_3 reacts with W to WO_2 which sublimes at $800\,^{\circ}\mathrm{C}$, far below the melting point of tungsten $T_m = 3410\,^{\circ}\mathrm{C}$ [112]. Additional benefits of flashing are the desorption of contaminants, e.g. remains of etch solution/solvents, and/or down diffusion of such to the tip shaft.

Shaping In preparation of the experiments from Chap. 6 and 7, the flashed tungsten tip was additionally dipped into a Au(111) crystal in a controlled manner. Soft-dipping the tip few nm into a good metal like gold in order to pick up some gold atoms is a widely used technique to modify the shape of the tip end. The procedure is repeated until a good lateral resolution and a relatively flat tip DOS is obtained. The fact that the d-orbitals of Au are completely filled provides a tip with mostly s-wave like states around the Fermi energy and the position of the d-states with respect to the Fermi energy also makes Au the noblest of all metals [113]. The former trait provides good accordance with quantum tunnelling theory in the Tersoff-Hamann approximation and a high tunnelling conductivity but an overall lower lateral resolution at small tip-sample distances due to the large radial extension of s-wave states. The latter is important for the experiments on Pb(111), described in Chap. 6 and 7 because a pure tungsten tip has higher attractive forces towards the Pb surface atoms than a gold covered tip. A bare tungsten tip often pulls out chains of Pb atoms or even picks them up during a scan. This could lead to the destruction of a flat surface area or a superconducting tip, which should both be avoided.

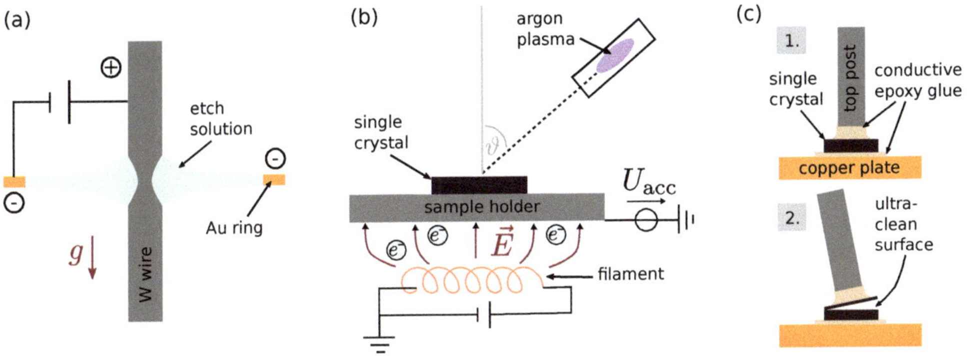

Figure 3.4: Tip and Sample Preparation: (a) Sketch of the experimental setup for electrochemical tip etching using the DC drop-off method. (b) Schematic depiction of the sputtering and annealing process. During sputtering, an ammeter between sample and ground is installed instead of the high voltage source (U_{acc}) that is present during annealing by electron bombardment. (c) Cartoon of the in-situ cleaving process using a top-post. The sample and top-post are glued (1) under ambient conditions. The cleave-off (2) is done in UHV at $\sim 88\,\mathrm{K}$.

3.3.2 Sputtering and Annealing

Sputtering A clean and smooth single crystal surface for the used metals Au(111)[2] and Pb(111)[3] was obtained by several cycles of argon sputtering and subsequent annealing. During the sputtering process, the single crystal is exposed to an argon ion beam of $U_{\mathrm{acc}} = 4\,\mathrm{kV}$ accelerating voltage at an angle of $\vartheta = 50°$ to the surface normal (see Fig. 3.4(b)). The argon gas[4] is let into the preparation chamber through a needle valve up to a pressure of $p_{\mathrm{prep}} \sim 2 \times 10^{-8}\,\mathrm{mbar}$ (base pressure is $p_{\mathrm{prep}} \sim 1 \times 10^{-10}\,\mathrm{mbar}$). The pressures can be kept relatively low since the sputter gun is differentially pumped by the load-lock. In the sputter gun, the sputter gas is ionized by free electrons emitted from a filament running at $I_{\mathrm{em}} = 8\,\mathrm{mA}$ emission current. With these parameters, the ion current between sample and machine ground is $I_{\mathrm{ion}} \sim 4.5\,\mu\mathrm{A}$. Since the raster scan of the sputter gun covers a $7 \times 7\,\mathrm{mm}^2$ area and not only the sample but also other parts of the manipulator head are connected to ground, this value is to be taken with caution when its compared to a different apparatus. It here refers to the experimental setup of the D-STM's preparation chamber. The high-energy argon ions transfer their energy and momentum to the sample's surface atoms leading to the ejection of single surface atoms or clusters.

[2] Single crystal Au(111) (miscut angle: $\pm 0.1°$, purity: 99.999% , hat shape, diameter: $6.9\,\mathrm{mm}$, thickness: $3\,\mathrm{mm}$) by MaTecK GmbH

[3] Single crystal Pb(111) (miscut angle: $\pm 0.1°$, purity: 99.999%, hat shape, diameter: $8\,\mathrm{mm}$, thickness: $2\,\mathrm{mm}$) by MaTecK GmbH

[4] Argon minican 5.0 (purity$\geq 99.999\,\%$) by Linde

As a result, the crystal surface is cleaned from surface contaminants but left with larger roughness than the initial polished crystal and some argon inclusions. With the parameters mentioned, roughly $1\,\mathrm{ML/min}$ is removed from the surface. Typical durations of a sputtering procedure were $10 - 20\,\mathrm{min}$.

Annealing In a subsequent annealing step, the sample is heated to a temperature below its melting point. The annealing procedure serves two purposes: It drastically enhances surface diffusion to smoothen the surface and induces the surface segregation of impurities. These impurities can then be removed in a consecutive sputtering procedure. The sample crystal was heated using the thermal radiation of a filament placed beneath it. For annealing temperatures above $300\,^{\circ}\mathrm{C}$, the sample was additionally heated by electron bombardment (see Fig. 3.4(b)). The crystal temperature can be monitored by an infrared pyrometer[5] and the manipulator temperature by a thermocouple. The Au(111) crystal was annealed at $T_{\mathrm{pyro}} \sim 550\,^{\circ}\mathrm{C}$. The heating protocol was 2 min 45 sec of electron bombardment at $I_{\mathrm{em}} = 26\,\mathrm{mA}$ and $U_{\mathrm{acc}} = 300\,\mathrm{V}$. The Pb(111) crystal was annealed at up to $T_{\mathrm{pyro}} = 260 - 280\,^{\circ}\mathrm{C}$ without high voltage for 6 min.

Hot Sputtering A combined process, called hot sputtering, was used for Pb(111). The simultaneous soft annealing at $T_{\mathrm{pyro}} < 200\,^{\circ}\mathrm{C}$ and sputtering can help to keep the sputtering damage low while constantly segregating contaminants to the surface.

3.3.3 In-situ Cleaving

For the cuprates used in this work, several layers were cleaved off in-situ to expose a fresh, ultra-clean surface. Due to their layered structure they can easily be cleaved in the c-plane by the top-post cleaving method shown in Fig. 3.4(c). The sample is with one side glued to the sample plate and with the other side to a metal post on top of it. The adhesive is a conducting epoxy glue[6]. As a result, the crystal is mechanically clamped (and therefore strained after a cooldown) and the tunnelling current can flow through the conductive glue which ensures a good electrical contact. The samples were cleaved off on the precooling stage in the STM chamber of the respective machine at $\sim 88\,\mathrm{K}$. The top-post was kicked off using a wobble-stick manipulator.

[5] OPTCTL3MH2 by Optris
[6] EPO-TEK H20E by EPOTEK

4 Bosonic Excitation Spectrum in $Bi_2Sr_2CaCu_2O_{8+\delta}$ and $YBa_2Cu_3O_{6+x}$

From a technological viewpoint, high temperature superconductors (HTS) represent the most desirable class of unconventional superconductors. Among them, the cuprates, which are layered copper oxides with perovskite structure, are perhaps the most famous representatives. The discovery of the first HTS lanthanum barium copper oxide (LBCO) [7] was awarded with a Nobel Prize in 1987 and soon after, other cuprate superconductors with transition temperatures above liquid nitrogen temperature followed. Prime examples are the widely studied systems of yttrium barium copper oxide (YBCO) [114] and bismuth strontium calcium copper oxide (BSCCO) [115]. In this chapter, we address the open question of what the bosonic glue in the cuprates is. After introducing the studied materials $Bi_2Sr_2CaCu_2O_{8+\delta}$ (Bi2212) and $YBa_2Cu_3O_{6+x}$ (Y123) and giving a brief overview of the scientific context, inelastic scanning tunnelling data is tested to tunnelling theory based on the spin-fermion model of spin-fluctuation mediated superconductivity.

4.1 Structural and Electronic Properties

4.1.1 Structural Properties of Y123 and Bi2212

The high-temperature superconducting phase of Bi2212 crystallizes in an orthorhombic $Amaa$ structure (point group D_{2h}). The unit cell is displayed in Fig. 4.1(a) with lattice constants $a = 5.41$ Å, $b = 5.40$ Å and $c = 30.72$ Å. This orthorhombic unit cell contains each of its elements four times its stoichiometric value. The slight difference in a and b supposedly results from the formation of a superlattice in the Bi-O planes [118]. Along the c-axis CuO_2 planes are sandwiched between SrO and Ca planes. The Ca planes are mirror planes of the unit cell. The SrO-CuO_2-Ca-CuO_2-SrO stacks are then again sandwiched by BiO planes. The BiO planes provide only weak bonding to each other. Careful micromechanical cleaving of this crystal preferably leads to a separation between the BiO planes due to their weak van-der-Waals bonds. The

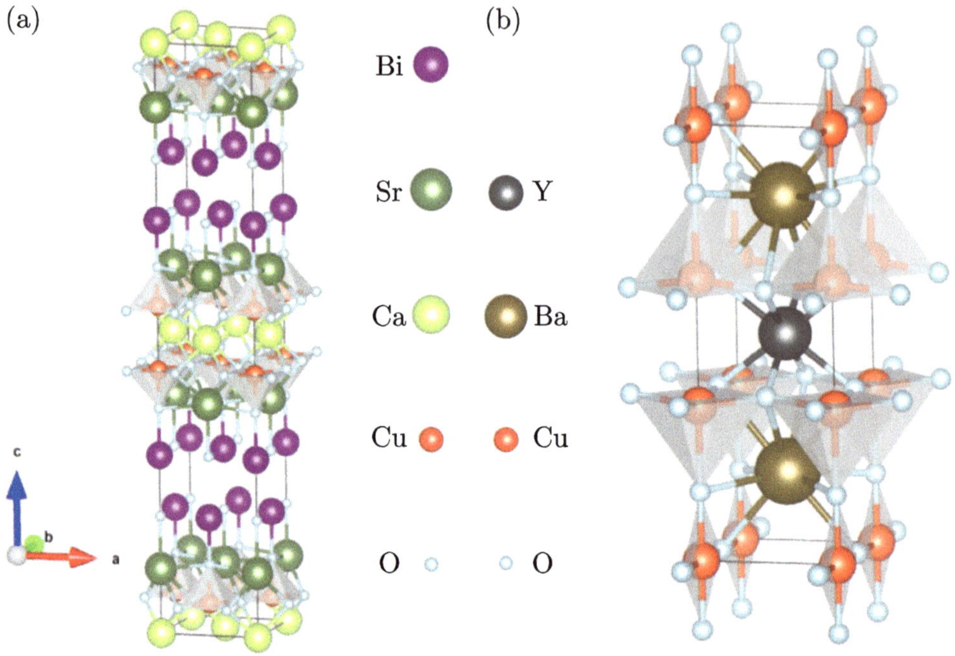

Figure 4.1: Structure Models: Crystal structures of (a) $Bi_2Sr_2CaCu_2O_{8+\delta}$ in the orthorhombic phase (space group *Amaa*) (b) $YBa_2Cu_3O_{6+x}$ in the orthorhombic phase (space group *Pmmm*). Models were created using crystallographic data from Ref. [116, 117].

appearance of other termination layers along the c-axis is possible but occurs in rarer cases [119].

The high-temperature superconducting phase of Y123 crystallizes in an orthorhombic $Pmmm$ structure (point group D_{2h}). The unit cell is displayed in Fig. 4.1(b) with lattice constants $a = 3.82$ Å, $b = 3.89$ Å and $c = 11.68$ Å. Starting from the perovskite structure of $YBa_2Cu_3O_9$, one oxygen atom is removed in every third CuO_2 plane and one oxygen atom is removed in the Y-plane. The result can be viewed as a layered unit cell that contains two CuO_2 planes separated by a Y-plane and one plane of CuO chains along the b-axis which are separated from the CuO_2 planes by BaO planes. The composite with low oxygen concentration $x < 0.35$ acquires tetragonal symmetry without CuO chains. The structural transition between the tetragonal and orthorhombic phase takes place inside the range $x = 0.35 - 0.65$ [118]. Unlike Bi2212, Y123 has no natural cleaving plane. The experimentally found predominant termination layers after cleaving the crystal are the BaO and CuO chain layers [120].

4.1.2 Unconventional Pairing in Y123 and Bi2212

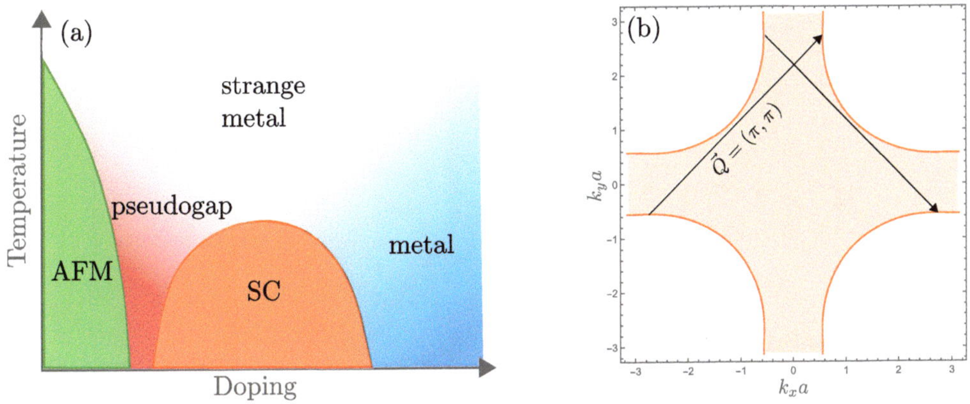

Figure 4.2: Electronic Properties: (a) Schematic phase diagram of hole-doped cuprate superconductors. Adapted from [22]. (b) Typical 2D Fermi surface contour of the hyperbolic bands in the copper oxide plane of cuprate superconductors including the AFV $\boldsymbol{Q}$ that connects hot spots on the Fermi surface.

Undoped cuprates tend to be antiferromagnetic insulators at low temperatures. In Bi2212 and Y123, sufficient hole doping, by regulating the oxygen concentration during growth, can destroy the magnetic order through the injection of charge carriers and reveal a rich variety of quantum matter phases. In Fig. 4.2(a), a simplified phase diagram for the hole-doped superconductors in the temperature-doping plane is shown. A superconducting dome with parabolic $T_c(p)$ dependence, where p denotes the number of holes per Cu atom, is found to be quite universal for cuprate superconductors [122–124]. The oxygen concentration for which T_c is maximal is called optimal doping, lower (higher) doped samples are considered underdoped (overdoped). For the hole doped double-layer HTS like Y123 and Bi2212, the dome is asymmetric with a steep decline of T_c in the overdoped regime when the hole doping is assumed to correspond to chemical composition ($p \approx x$). A possible explanation for this trend is a redistribution of oxygen in the overdoped regime [124]. The parabolic dome is, however, recovered if the hole doping is inferred from the Fermi surface volume in angle-resolved photoemission spectroscopy (ARPES) measurements [125]. For Bi2212, the dome is centred around $p = 0.18$ with $T_c^{\max} \approx 91\,\mathrm{K}$ [125], for Y123 $T_c^{\max} \approx 93\,\mathrm{K}$ is reached at $p \approx 0.17$ [126]. The most mysterious and least understood part of the phase diagram is the anomalous normal state (strange metal) at high temperatures that exhibits non-Fermi liquid behaviour[1]. This behaviour is

[1] A trademark of this anomalous normal state is a resistivity that scales linearly instead of quadratically (Fermi-liquid) with temperature.

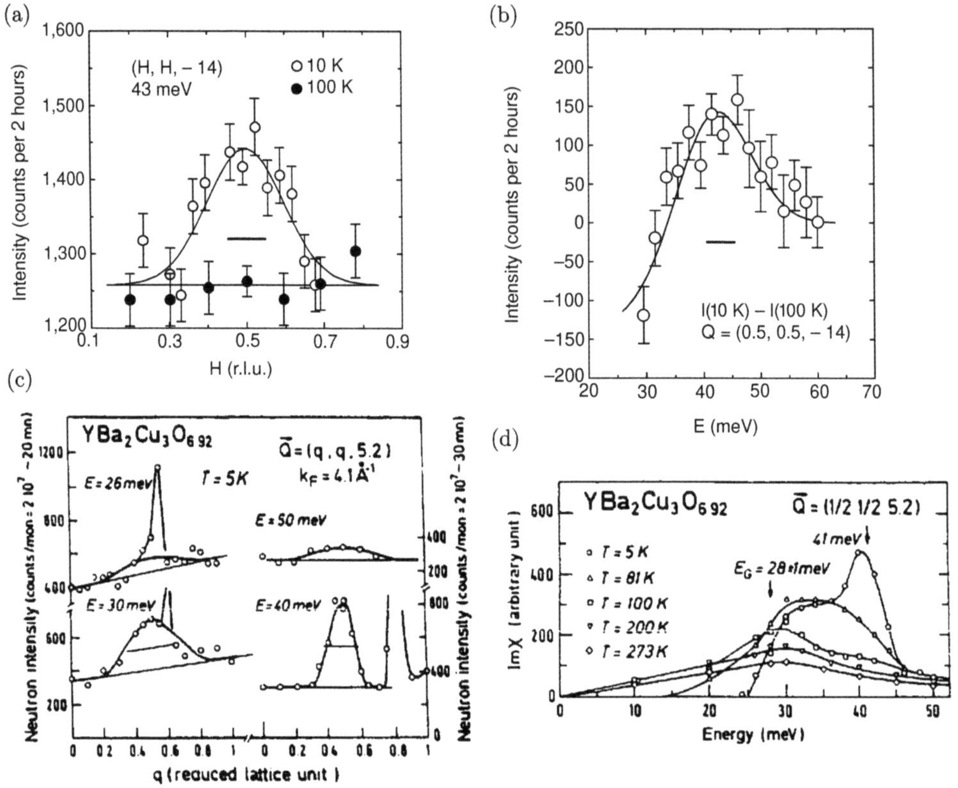

Figure 4.3: Spin Excitation in INS: Inelastic neutron scattering data of optimally doped Bi2212 (a-b) and Y123 (c-d): (a,c) $\boldsymbol{Q}$ scans at the indicated temperatures revealing the resonance mode at the AFV (π,π,l). (b,d) Energy scans at the AFV showing the dispersion of the spin excitation at $\omega_{\text{res}} \approx 43\,\text{meV}$ for Bi2212 and $\omega_{\text{res}} \approx 41\,\text{meV}$ for Y123. For low excitation energies both materials show a spin gap. (a,b) are reprinted by permission from Springer Nature Customer Service GmbH: Springer Nature [26], Copyright (1999). (c,d) are reprinted from [121], Copyright (1991), with permission from Elsevier.

most pronounced in the underdoped cuprates at elevated temperatures between the superconducting and magnetically ordered phase, where the material enters a so-called pseudogap phase below the temperature T^*. The name is derived from the opening of a gap in the electronic spectrum that can be seen in ARPES or STM measurements. If this phase is closely tied to spin fluctuations and represents preformed Cooper pairs [127] or if it is the result of a competing order unrelated to superconductivity, like nematicity [126], remains to be clarified.

The relatively large c/a ratio in the cuprates leads to an effective two-dimensional electronic bandstructure lying in the 2D Brillouin zone (BZ) of the copper-oxide planes which are separated by ionic spacer regions. As a result, several electronic properties of the normal, but also the superconducting state, like conductivity, superconducting coherence length or magnetic penetration depth have large out-of-plane anisotropy [56]. The strong confinement of charge carriers to the copper-oxide planes gives rise to the strong doping dependence of electronic properties and the lowered dimensionality favours quantum-criticality [22, 28]. Fig. 4.2(b) shows the generic Fermi surface contour (orange line) stemming from the square CuO_2 plane in a cuprate superconductor in the 2D BZ. The shaded area around the Γ point represents the occupied states and the white area round the M points the unoccupied states. At zero doping ($\delta = 0$) the percentage area of the two is identical. In the case of hole doping ($\delta > 0$) the relative area of unoccupied states in the first BZ dominates.

Nuclear magnetic resonance (NMR) experiments in the early 1990s could confirm a spin-singlet state in the cuprates [128] and thus a superconducting state with an even parity orbital wavefunction. As was already hinted at in Sec. 2.5.1, the famous representatives of the cuprate family Y123 and Bi2212 are in an orthorhombic D_{2h} phase at temperatures close to T_c and the relevant doping rage. Hence, for the irrep A_{1g}, a superposition of extended s-wave ($s + s_{\pm} + ..$) and $d_{x^2-y^2}$ basis functions is technically allowed. A pairing symmetry based on the other irreps is unlikely considering the nearest-neighbour and next-nearest neighbour positions on the rectangular CuO_2 lattice to which the conduction is restricted. Due to the small orthorhombicity in these systems, the superconductivity is widely believed to arise from an unconventional spin-fluctuation pairing mechanism of predominantly $d_{x^2-y^2}$ character in a near tetragonal D_{4h} phase[2]. There, such a $d_{x^2-y^2}$ pairing symmetry belongs to the B_{1g} irrep and consequently lowers the total symmetry in k-space. This order parameter symmetry is readily supported by numerous experimental evidence including results from thermodynamic measurements, ARPES, NMR, Raman scattering, SQUID interferometry and many more (see Ref. [56, 92] and references therein). An extensive summary of convincing high resolution ARPES data has been published by A. Damascelli and co-workers [129]. Still, the argument that an A_{1g} order parameter representation reveals no additional broken symmetry remains strong and consequently an $s + d$ pairing symmetry and even an involved conventional pairing mechanism are not completely out of the picture yet [56].

The immediate vicinity of an antiferromagnetic phase close to superconductivity hints towards strong magnetic fluctuations and strong short ranged spin correlations, i.e. an enhanced magnetic susceptibility (cf. Eq. (2.58)). On the (almost) square copper oxide planes, antiferromagnetic stripe order is realized for the AFV $\boldsymbol{Q} \approx (\pi, \pi)$. As shown in Fig. 4.2(b), this wave vector connects hot spots on the Fermi surface at

[2] This means the 2D symmetry group is C_{4v} in the copper-oxide plane instead of C_{2v} and so the CuO_2 unit cell is square instead of a rectangular one.

which the DOS is large. Even outside the magnetically ordered state, where the spin correlation length is smaller, the susceptibility has been shown to be maximal around $\boldsymbol{Q}$. In fact, INS experiments show evidence of a magnetic resonance mode at the AFV and opening of a spin-gap below T_c in nodal directions, as shown in Fig. 4.3[3]. The magnetic resonance is a collective spin-1 excitation mode in the superconducting state and is arising from the renormalization of spin-fluctuations through the gapped fermionic degrees of freedom, as could be demonstrated in the spin-fermion model (see Sec. 2.5.3). Experimentally, the resonance mode was found to be ubiquitous in the cuprate superconductor family and its energy is tied to T_c and Δ [130].

A spin-fluctuation mediated coupling mechanism in the spin-fermion model framework, as described in Sec. 2.5.3, logically combines the $d_{x^2-y^2}$ gap parameter with the hyperbolic 2D bands and magnetic susceptibility found in these materials. Since this theory is rather complicated, a simple and short answer to the question how $d_{x^2-y^2}$ pair symmetry and the AFV $\boldsymbol{Q} = (\pi/a, \pi/a)$ are related could sound as follows: Suppose an electron with momentum $\boldsymbol{q}$ hops from Cu lattice site $i+1$ to an adjacent Cu lattice site i on the square CuO_2 lattice. Then, it essentially carries the relative phase $\boldsymbol{q}a$ which is equal to π for $\boldsymbol{q} = \boldsymbol{Q}$. Now, the on-site Coulomb repulsion on a lattice site i drastically reduces the correlation for two electrons on the same site. It can, however, become smaller for electrons with opposite signed wavefunctions that stem from an $l > 0$ orbital because, there, these electrons are spatially separated. The electrons from the Cu $d_{x^2-y^2}$ orbitals provide this trait and hence allow for a non-zero correlation for electrons with opposite spin and relative wavevector $\boldsymbol{Q}$ at the same location $\boldsymbol{r}$.

4.2 Inelastic Tunnelling in Spin-Fluctuation Driven Superconductors

As stated in Sec. 2.5.3, in unconventional superconductors the integrated spin spectrum $g^2\chi''(\omega)$ (see Eq. (2.62)) plays the role of the Eliashberg function. Using the Keldysh diagrammatic formalism for Greens functions, P. Hlobil and J. Schmalian could show that this function is obtainable from the inelastic tunnelling conductance

[3] The wave vector $(\pi/a, \pi/a)$ corresponds to $h = k = 1/2$ in reciprocal lattice units (r.l.u.).

[28]. The inelastic tunnelling conductance due to coupling of electrons to spin fluctuations is given by

$$\sigma^{\mathrm{i}}(eV>0) = -\frac{\sigma_0}{D^2\nu_S^0}\int_{-\infty}^{\infty} \mathrm{d}\omega_1 \mathrm{d}\omega_2\, g^2\chi''(\omega_1)\tilde{\nu}_S(\omega_2)\times$$
$$\times\, n_F'(\omega_1+\omega_2-eV)[1-n_F(\omega_2)][1+n_B(\omega_1)]\,, \tag{4.1}$$

with the normalized sample DOS $\tilde{\nu}_S(\omega) = \nu_S(\omega)/\nu_S^0$, Fermi- and Bose-Einstein distribution functions n_F and n_B, as well as the cut-off energy for the bosonic excitation spectrum D. For a two-dimensional Brillouin zone, Eq. (2.62) in the normal state simplifies to

$$\chi''(\omega) \propto \nu_S^0 \arctan(\omega/\omega_{sf}). \tag{4.2}$$

This leads to $\sigma^{\mathrm{i}} \propto g^2(eV)^2/\omega_{sf}^2$ for small voltages $eV \ll \omega_{sf}$ and $\sigma^{\mathrm{i}} \propto g^2 eV/\omega_{sf}$ for $eV \gg \omega_{sf}$. This behaviour in the normal state is shown in the blue curve of Fig. 4.4(b), which shows the inelastic tunnelling conductance derived from the formal convolution of a constant normal state DOS (blue curve in Fig. 4.4(a)) and an overdamped spin spectrum following Eq. (4.2). In the superconducting state, the spin spectrum is renormalized and a spin gap below ω_{res} evolves (see Sec. 2.5.3). Performing the formal convolution of the superconducting DOS (red curve in Fig. 4.4(a)) and the integrated spin spectrum in the superconducting state (Fig. 2.8(b)), as instructed in Eq. (4.1), one finds the total tunnelling conductances $\sigma^{\mathrm{tot}} = \sigma^{\mathrm{el}} + \sigma^{\mathrm{i}}$ shown in Fig. 4.4(c) for different weights $1/D$ of the inelastic tunnelling contribution. While the pure elastic tunnelling conductance shows a peak-like feature at $\Delta + \omega_{\mathrm{res}}$

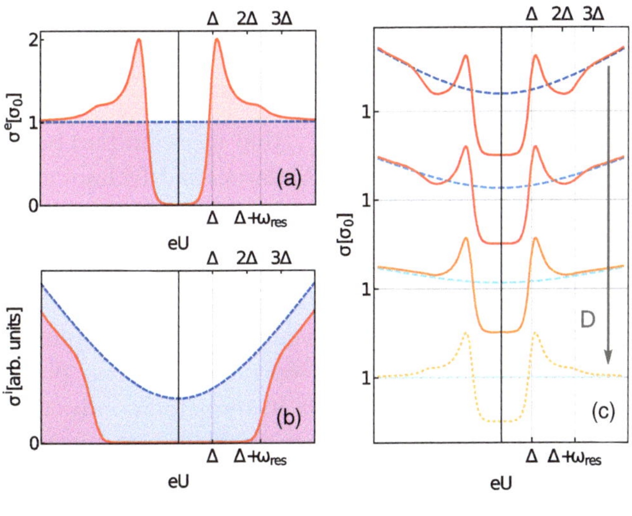

Figure 4.4: Elastic and Inelastic Contributions: Calculated elastic (a) and inelastic (b) contributions to the differential conductance in superconducting (red) and normal state (blue). The total conductance (c) develops a dip-hump structure around $\Delta+\omega_{\mathrm{res}}$ for high inelastic current contributions (red curves). For small to zero inelastic current contributions (orange/yellow curves) the conductance in the superconducting state stays higher than the conductance in the normal state for $eU > \Delta$. Reprinted figure with permission from [28]. Copyright (2017) by the American Physical Society.

due to coupling to virtual bosons, the inclusion of inelastic tunnelling events leads to a peak-dip-hump structure in the total conductance for energies larger than Δ. The hump at $\sim \Delta + \omega_{\text{res}}$ that brings the total conductance in the superconducting state back to match the one in the normal state is due to the sudden onset of inelastic tunnelling events involving real bosonic excitations.

From a series of INS experiments on various cuprate superconductors, the universal relationship between magnetic resonance and superconducting gap of

$$\frac{\omega_{\text{res}}}{\Delta} = (1.28 \pm 0.08) \tag{4.3}$$

was found [130]. While the resonance mode is directly obtained in INS, more model assumptions, like the applicability of the spin-fermion model, have to be put in, to deduce the spin excitation spectrum from STS and in addition to that, the k-space information is lost in tunnelling experiments. In previous attempts at modelling the Eliashberg function, *a priori* knowledge on its shape had to be put in, which biases the result tremendously due to the ill-posedness of the deconvolution problem [131]. Recently, the application of machine learning algorithms on ARPES data proved to be a powerful concept to reverse-model the spin-spectrum, but this happens at the cost of a number of free parameters which cannot be easily mapped onto physical quantities [132, 133]. Instead, we want to pursue the path of a direct deconvolution of scanning tunnelling data, using *a priori* band structure for the normal state model and the inelastic tunnelling theory described above, as input. That this yields insightful results, has already been demonstrated for monolayer FeSe [27]. The goal here is to extend this deconvolution method to the nodal cuprate superconductors.

4.3 Separation of Elastic and Inelastic Tunnelling Events

The single crystals $Bi_2Sr_2CaCu_2O_{8+\delta}$ and $YBa_2Cu_3O_{6+x}$, that were studied here, were grown by Thomas Wolf at the Institute for Quantum Materials and Technologies (IQMT). The Bi2212 sample is slightly underdoped with a T_c of 82 K. The Y123 sample is from the batch WAX350−1 and is optimally doped with a T_c of 92 K.

In search of the bosonic mode at $\Delta + \omega_{\text{res}}$ in low-temperature scanning tunnelling spectra, it is first important to take a look at the inhomogeneity of the samples to decide which spectra are most promising to study. As reported by Fischer et al., Bi2212 tends to show a large inhomegeneity of its DOS in the superconducting state [120]. This can be confirmed in the direct comparison of the conductance inhomegeneity measured on Bi2212 and Y123, shown in Fig. 4.5. The heat maps of the conductance variation $\delta\sigma/\bar{\sigma}(x,y) = \sum_{-eU_t}^{eU_t}(\sigma(eU,x,y) - \bar{\sigma}(eU))/\bar{\sigma}(eU)$ show that it

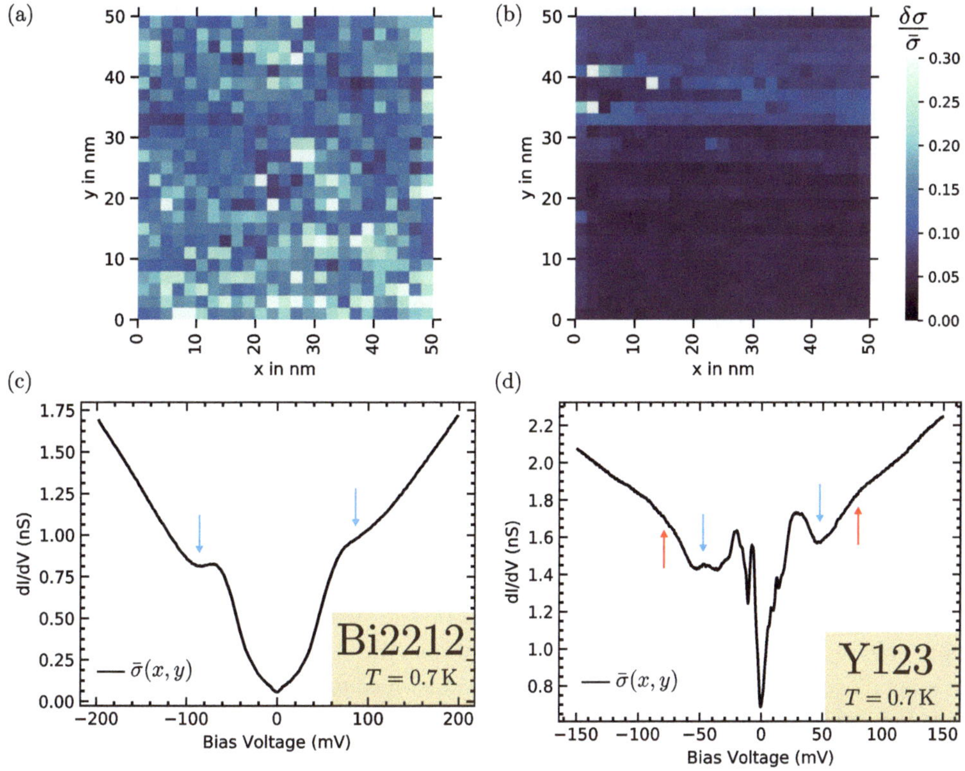

Figure 4.5: Conductance Inhomogeneity: (a,b) Heat maps showing the variation of the differential conductance over a $50 \times 50\,\mathrm{nm}^2$ surface area on Bi2212 (a) and Y123 (b). (c,d) Position averaged bias spectra on the grids shown in (a,b) at $T = 0.7\,\mathrm{K}$. The higher inhomegeneity of the Bi2212 surface is reflected in both the conductance variation map and the blurred position averaged spectrum. The characteristic dip and hump are marked by blue and red arrows.

is with $\approx 30\,\%$ on Bi2212 about three times higher than on Y123. As a consequence, a position averaged spectrum over a $50 \times 50\,\mathrm{nm}^2$ area can preserve detailed gap features better for Y123 (Fig. 4.5(d)) than for Bi2212 (Fig. 4.5(c)). Especially the dip-hump (dip marked by blue, hump marked by red arrows in Fig. 4.6(c,d)) feature is still clearly visible in the position averaged spectrum of Y123 at $\epsilon \approx 60\,\mathrm{meV}$ but is invisible in Bi2212. Therefore, in the case of Bi2212, an average spectrum at one specific location, at which the dip-hump spectral feature was clearly visible, was chosen for this study. For Y123, the position averaged spectrum was chosen.

Fig. 4.6 shows the measured differential conductance and the numerical derivative of the tunnelling conductance on the two sample surfaces. For Bi2212, the normal state at $84\,\mathrm{K}$ was measured with the same tip as the superconducting state at $0.7\,\mathrm{K}$. For Y123, only a spectrum in the superconducting state could be obtained because

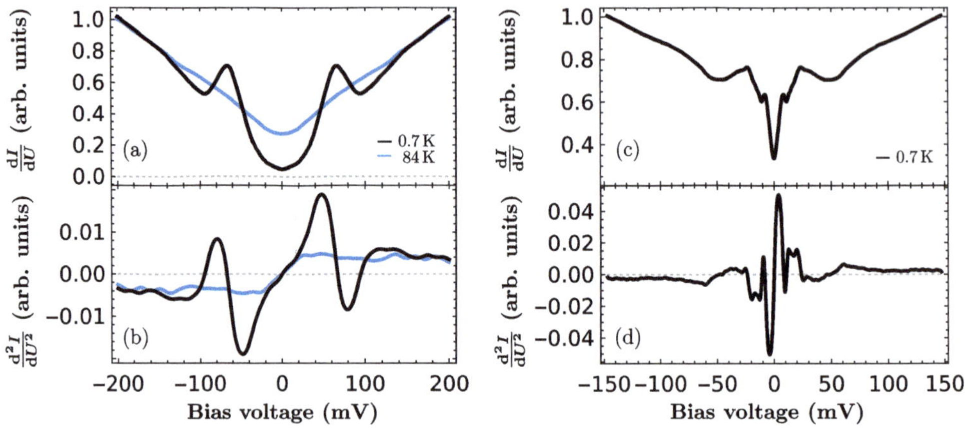

Figure 4.6: Symmetrized Spectra: First and second derivative of the tunnelling current on Bi2212 (a,b) and Y123 (c,d) in the normal (blue) and superconducting state (black).

drastic tip changes complicated the measurement at elevated temperatures. Due to the large inhomegeneity of the gap of Bi2212 on the surface, the spectrum presented here was recorded at one fixed position whereas the spectrum for Y123 is position averaged over an area of $50 \times 50\,\mathrm{nm}^2$. All spectra were symmetrized, normalized and regularized by Gaussian filtering. The symmetrization accounts for the asymmetric background, seen in the spectra (see Fig. 4.5 (c,d)). Most tunnelling experiments so far, revealed a tilted background conductance that is higher in the negative than in the positive bias range due to the influence of doping on the normal state DOS of Bi2212 and Y123 [120]. Since the tilt here is in the other direction, it is most likely caused by a non-constant tip DOS in the applied bias voltage range. The normalization is chosen such that conductance in the normal and superconducting state are the same for $E \gg \Delta$. This enables a better comparison with theory, where the normal state is effectively studied at 0 K. In reality, the background conductance is reduced at high temperatures due to increased electron-phonon scattering. The regularization smooths the $\mathrm{d}I/\mathrm{d}U$ curves for a better display of the numerical derivative and prepares it for the deconvolution procedure in the following section.

The spectrum for superconducting Bi2212 in Fig. 4.6(a) shows a single gap with remanent zero-bias conductance due to the $d_{x^2-y^2}$ gap symmetry and relatively smeared coherence peaks hinting at short quasiparticle lifetimes. This is typical for the underdoped regime and may be caused by its proximity to the insulating phase [120]. Outside the gap, a clear dip of the superconducting spectrum below the normal conducting spectrum, followed by a hump reapproaching it, are visible. The V-shaped conductance in the normal state hints towards strong inelastic contributions to the tunnelling current, as was demonstrated in Fig. 4.4. The hump shows as a peak in the second derivative of the tunnelling current that exceeds the curve of the normal state at $\approx 130\,\mathrm{mV}$ in Fig. 4.6(b). The relatively round shape of the superconducting gap in Bi2212 is atypical for a classic *d*-wave superconductor, in which the naive expectation is a V-shaped conductance minimum. As will be shown, the round shape of the gap can, however, be generated without admixture of an *s*-wave paring term by respecting the anisotropy of the Fermi surface in the normal state.

The spectrum for superconducting Y123 in Fig. 4.6(c) is qualitatively in excellent agreement with previous STM measurements [134] and shows three low-energy features: (i) a superconducting coherence peak at $\approx 25\,\mathrm{meV}$ that is less broadened than in Bi2212, (ii) a high-energy shoulder of the coherence peak and (iii) a low-energy peak at $\approx 10\,\mathrm{meV}$. A good explanation for the high-energy shoulder and sub-gap peak could be that they arise from the proximity-induced superconductivity in BaO planes and CuO chains [135, 136]. This would certainly fit the fact that these states are missing in the Bi-based compounds and that the sub-gap peak shows a direction-dependent dispersion in ARPES data [129]. The hump lies at $\approx 60\,\mathrm{meV}$. Even without a normal conducting spectrum, the V-shaped background conductance, which is in agreement with the predicted inelastic contribution by magnetic scattering in the spin-fermion model, is obvious in the high-energy regime of the superconducting spectrum.

In order to extract the inelastic part of the tunneling spectrum, it is necessary to determine the elastic contribution by fitting a model function that keeps the complexity as low as possible while still capturing the relevant features of the band structure and pairing strength. To account for the presence of pair-breaking processes, a generalized Dynes function was chosen [137]:

$$\begin{aligned} \nu_s^{\mathrm{el}}(\omega) &= \frac{\sigma_0}{n}\int_0^{\pi/2} \mathrm{d}\varphi \nu_0 \nu_{\mathrm{n}}^{\mathrm{F}}(\varphi) \times \\ &\times \left| \mathcal{R}\left(\frac{\omega + i\Gamma(\varphi)}{\sqrt{(\omega + i\Gamma(\varphi))^2 - \Delta(\varphi)^2}} \right) \right| . \end{aligned} \tag{4.4}$$

Here, $n = \int \mathrm{d}\varphi \nu_{\mathrm{n}}(k_F, \varphi)$ is a normalization factor, $\Delta(\varphi) = \Delta_0 \cos(2\varphi)$ is the d-wave pairing potential, $\Gamma(\varphi) = \gamma \cdot |\Delta(\varphi)|$ the scattering rate and $\nu_{\mathrm{n}}^{\mathrm{F}}(\varphi)$ is the angle dependent density of states at the Fermi energy, in the normal conducting phase. The function $\nu_{\mathrm{n}}^{\mathrm{F}}(\varphi)$ weighs gap distributions for different (k_x, k_y) by their abundance along the Fermi surface. In order to determine this function, the in plane dispersion is modeled by a tight binding model of the form

$$\begin{aligned}\epsilon(k_x,k_y) = &\frac{t_1}{2}(\cos(k_x)+\cos(k_y)) + t_2\cos(k_x)\cos(k_y)\\ &+\frac{t_3}{2}(\cos(2k_x)+\cos(2k_y)) + \frac{t_4}{2}(\cos(2k_x)\cos(k_y)\\ &+\cos(k_x)\cos(2k_y)) + t_5\cos(2k_x)\cos(2k_y)\\ &-\mu,\end{aligned} \tag{4.5}$$

with chemical potential μ and hopping parameters t as proposed in Ref. [138] for a near optimally doped Bi2212 crystal. For Y123, the model function [139]

$$\epsilon(k_x,k_y) = 2t_1\cos(k_y)2t_3(\cos(2k_x)+\cos(2k_y)) \tag{4.6}$$

$$\pm\left[4\cos^2(k_x)(t_1+2t_2\cos(k_y))^2+V^2/4\right]^{1/2}+\mu \tag{4.7}$$

was used, where the fourth term accounts for the ortho-II band folding due to CuO chains. The parameters $t_1 = -0.558\,\mathrm{eV}$, $t_2 = -0.31t_1$, $t_3 = -0.5t_2$, $\mu = 0.17\,\mathrm{eV}$ and $V = 0.025\,\mathrm{eV}$ were used. The parameters μ and V were adjusted such that the three resonances in the DOS from the quasi-1D band correspond to the energetic positions of peaks in the spectrum. Compared to Ref. [139], the lower μ corresponds to lower hole-doping and shifts the lowest energy resonance to lower energies compared to the high energy resonance (coherence peak shoulder). A reduced ortho-II potential V leads to a more pronounced low-energy resonance due to the LDOS peak in the near-nodal direction as seen in Fig. 4.7 (d).

An analytic expression for the Fermi wave vector $k_F(\varphi)$ is retrieved from the solution of $\epsilon(k,\varphi) - \mu = 0$ where $\epsilon(k,\varphi)$ is the polar representation of Eq. (4.5). The normal DOS is then given by

$$\nu_\mathrm{n}^\mathrm{F} = \oint \frac{\mathrm{d}^2k}{|\nabla_{\boldsymbol{k}}\epsilon|} \to \int_l \frac{\mathrm{d}\varphi}{\nabla_{k,\varphi}\epsilon(l(\varphi))}\,\|l'(\varphi)\|_2\,, \quad \varphi\in M_\varphi, \tag{4.8}$$

where $l(\varphi) = (k_F(\varphi)\cos(\varphi), k_F(\varphi)\sin(\varphi))^T$ is a parametrization of the path along the Fermi surface, $\|\cdot\|_2$ is the Euclidean norm and $M_\varphi = \{\varphi | k_F(\varphi) \in 1.\mathrm{BZ}\}$.

The calculated Fermi surface contour and DOS in the normal state are shown in Fig. 4.7. The fits to the superconducting spectra using Eq. (4.4) and parameters listed in Tab. 4.1 are shown in Fig 4.8(a,b). Unlike in fully gapped superconductors, the angle-integrated inelastic spectrum can be non-zero down to vanishing bias voltage because $\Delta(\boldsymbol{k})$ has a nodal structure. This prevents a direct assignment of the differential conductivity for $e|U| \lesssim \Delta$ to the purely elastic tunneling contribution like it is possible for the $s_\pm$ superconductor monolayer FeSe [27, 28]. Instead, the fact that

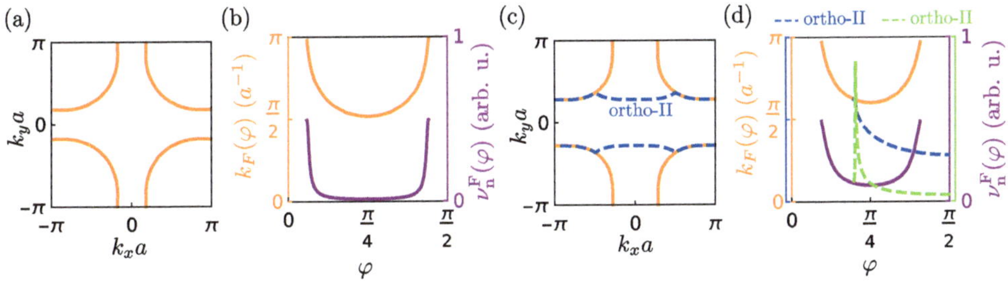

Figure 4.7: Normal State Electrons: (a,c) Calculated Fermi surface in the first 2D BZ of Bi2212 (a) and Y123 (c). For Y123 the contribution of the quasi-1D ortho-II band is shown in a dashed blue line. (b,d) Calculated Fermi wavevector $k_F(\varphi)$ (orange) and normal conducting DOS along the Fermi surface contour $\nu_n^F(\varphi)$ (purple) as function of polar angle in the first BZ quadrant for Bi2212 (b) and Y123 (d). For the ortho-II band, the Fermi wavevector and normal DOS are drawn as blue and green dashed lines in (d).

$\sigma^{\text{tot}} = \sigma^{\text{el}} + \sigma^{\text{inel}}$ and the physical constraints $\sigma^{\text{inel}}(0) = 0$ and $\sigma^{\text{inel}}(e|U| > 0) > 0$, shall be used. As demonstrated in Sec. 4.2, σ^{inel} is in principle a convolution of ν_s and the bosonic spectral function. Hence, for a slowly varying bosonic function, which is expected in the range $0 < e|U| < \Delta$, σ^{inel} is dominated by the elastic contribution $\nu_s(\omega)$ with some factor, say $(1-\alpha)$. Consequently, in the range $0 < e|U| < \Delta$, σ^{el} is well guessed by our Dynes fit times a factor $\alpha < 1$. For Bi2212 α is chosen in the following way: From the requirement that the integrated elastic contribution to the conductance in the spectroscopic range is equal for normal and superconducting state, one can find the scaling factor for the elastic part that best describes its proportion to inelastic processes. This is shown in Fig. 4.9(a) in Sec. 4.4. For Bi2212 this proportionality factor is $\alpha = 0.6$. In order to keep the condition $\sigma^{\text{el}}(\infty) = \sigma_0$, the experimental curve is scaled up by $1/\alpha$ instead of scaling down the fitted curve. The resulting elastic and inelastic contributions are shown in red/grey in Fig. 4.8(c,d). For Y123, only the condition $\sigma^{\text{inel}} > 0$ was enforced by using the scaling factor $\alpha = 0.9$ because normal state spectra were lacking. The influence of this scaling factor on the extracted bosonic spectrum is discussed in detail in the next section.

Table 4.1: Fit Parameters: Fitted Dynes function (Eq. (4.4)) parameters for Y123 and Bi2212 superconducting spectra.

	Δ_0 (meV)	ν_0 (σ_0)	γ
Bi2212	63.3	0.47	0.15
Y123	29.53	0.083	0.08
Y123-oII	19.89	0.610	0.58

4.4 Determining the Dispersion of the Bosonic Glue

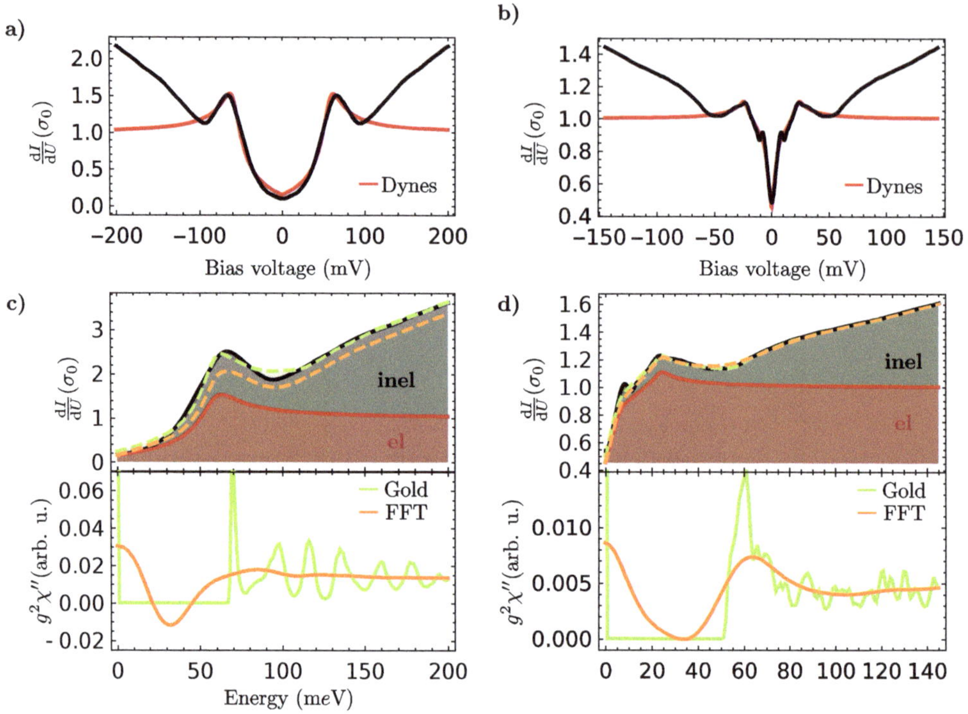

Figure 4.8: Spin Spectrum Extraction: (a,b) Elastic conductance fit for $eV \lesssim \Delta$(red) following Eq. (4.4) with parameters from Tab. 4.1 to the measured conductance on Bi2212 (a) and Y123 (b). (c,d) Top: The measured conductance spectrum is scaled up by $1/\alpha$. The inelastic part is the difference between total conductance (black line) and the elastic part (red line). Red and orange dashed lines show the total conductance obtained from convolving the elastic conductance guess with the orange and green spin spectra. Bottom: Spin spectra obtained from direct deconvolution in Fourier space (orange) and from the Gold deconvolution algorithm (green). Results for Bi2212 are shown in (c), for Y123 in (d).

In order to extract the spin spectrum, Eq. (4.1) at $T = 0$ is rewritten as a formal convolution integral:

$$\begin{aligned}\sigma^{\text{inel}}(eU) &\propto \int_0^{\text{eU}} d\omega \nu_s(\omega) g^2 \chi''(eU-\omega) \\ &= \left([\nu_s \cdot \Theta] * \left[g^2\chi'' \cdot \Theta\right]\right)(eU)\,. \end{aligned} \tag{4.9}$$

As can be seen from Eq. (4.9), the integrated spin spectrum χ'' is in information processing terms the source, the DOS in the superconducting state is the kernel and the inelastic tunneling conductivity is the signal. Θ denotes the Heaviside step function. In general, retrieving the function $g^2\chi''$ from the inelastic conductivity is an ill-posed problem. Additional aspects that complicate the problem are the following:

1. The kernel function $\nu_s(\omega)$ cannot be determined with enough certainty. As a result, we are on the verge of a necessity for blind deconvolution algorithms.
2. Every involved function is nonzero for $U \to \infty$. However, for an efficient inversion of the problem the fast Fourier transform (FFT) method is unavoidable. Hence, all function arrays must be reasonably padded such that circular and linear convolution become similar.
3. Due to flaws of the fitted model and electronic noise, the inelastic tunneling conductivity is not monotonously rising. This can result in negative parts in the source spectrum which are unphysical.

We compare two methods with which the Eliashberg function was determined: By direct deconvolution in Fourier space and by the Gold algorithm [140, 141]. The advantage of the Gold algorithm is that for a positive kernel and signal, the result of this iterative deconvolution method is always positive, in agreement with our physical constraint. It is explained in more detail in Appendix A.

The regularized bosonic function from direct deconvolution in Fourier space

$$g^2\chi''(\omega > 0) = \mathcal{F}^{-1}\left(\frac{\mathcal{F}(\sigma^{\text{inel}})(t)}{A\sqrt{2\pi}\mathcal{F}(\nu_s \cdot \Theta)(t)}\right) \tag{4.10}$$

is shown in Fig. 4.8(c,d) in orange. The contributions at low energies mostly stem from the scaling factor α. The negative part of the spectrum for $E < \Delta$ is due to the abrupt change in elastic conductivity at zero energy (multiplication with Heaviside distribution) which results in heavy oscillations in time space. Therefore, the circular deconvolution contains nonzero contributions for $E < 0$ and negative contributions for $0 < E < \Delta$ which are not physical. Despite these shortcomings, the bosonic spectrum recovers well the tendency of the total conductivity (Fig. 4.8(c,d) dashed orange) and shows the expected behaviour at medium and high energies, i.e. a resonance at $\Delta < E < 2\Delta$ and approach of the normal state bosonic function for $E < 3\Delta$.

Using the result of the direct deconvolution method as a first guess to the Gold algorithm, the bosonic spectrum shown in green in Fig. 4.8(c,d) was obtained after 2,000,000 iterations. Again, the high value at $E = 0$ is a consequence of the scaling with factor α. Negative contributions are now gone and the bosonic function recovers well the total elastic conductivity shown in Fig. 4.8(c) (dashed green), especially the dip-hump structure, and shows a very clear resonance at $\omega_{\text{res}} \approx 70\,\text{meV} \approx 1.1\Delta_0 \approx$

$1.2\Delta_{max} \approx 1.5\bar{\Delta}$ for Bi2212 and at $\omega_{res} \approx 60\,meV \approx 2.0\Delta_0 \approx 2.4\Delta_{max} \approx 3.6\bar{\Delta}$ for Y123.

The resonance mode of Bi2212 extracted in this work is higher in energy than reported in inelastic neutron scattering experiments ($\omega_{res} \approx 43\,meV$ at the AFV) [26] and closer to the resonance determined by optical scattering ($\omega_{res} \approx 60\,meV$) [142]. Due to the loss of k-space information in tunnelling, the centre of the resonance is expected to be shifted to higher energies compared to the INS results (compare Fig. 2.8(a) and (b)). The ratio $\omega_{res}/\Delta_{max}$ lies within the current range of error of Eq. (4.3). In most other extraction methods of the bosonic mode energy, the normal state DOS is not respected, which is why the Δ_0 used there is most similar to what is here called Δ_{max}. Δ_{max} is the largest gap value that contributes to the elastic conductance spectrum.

For Y123, the resonance mode is significantly higher in energy in the bosonic density spectrum than experimentally found by INS in optimally doped samples with $\omega_{res} \approx 41 - 46\,meV$ [121, 130] and even lies outside the onset of the spin scattering continuum at $\omega_c \approx 59\,meV$ in the slightly underdoped regime [130]. Apart from the k-space integration, which shifts the peak centre to higher energies, several other factors can play a key role: (i) The well-studied 41 meV odd-parity mode is paired with an even-parity mode at $\omega^{e}_{res} \approx 53 - 55\,meV$ [130, 143–145] which may be of the same origin as it vanishes at T_c. This mode appears with a $\approx 3 - 20$ times lower intensity in INS than the odd-parity mode, but this does not necessarily have to hold for a tunnelling experiment. (ii) The bosonic spectrum extracted here is essentially poisoned by phononic contributions from every k-space angle. Even INS data at specific q-vectors have to be corrected by known phonon modes. In the integrated spectrum obtained here, this is simply not feasible anymore. (iii) Apart from physical arguments, there can also be made sceptical remarks on the

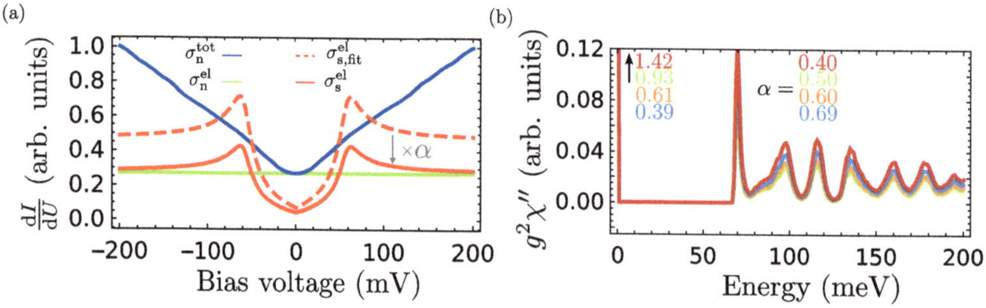

Figure 4.9: Numerical Scaling Factor α: (a) Determination of α happens through the boundary condition $\int d\epsilon\sigma_n^{el} = \int d\epsilon\sigma_s^{el}$. (b) Variation of α leaves the general shape of the extracted bosonic spectrum unaffected except for the magnitude of its zero-energy peak.

deconvolution procedure: Evidently, it heavily depends on the guess of the elastic tunnelling conductance, which in this case does not contain strong-coupling features from an Eliashberg theory, like in Fig. 4.4. The large deviation from $\omega_{res}/\Delta \approx 1.28$ is most likely caused by the determination of Δ. For gap sizes, the authors from Ref. [130] use a constant $\Delta_{max} = 39.5\,\mathrm{meV}$ throughout a large range of doping in the underdoped regime. In a collection of STS results spanning over 20 experiments, however, the extracted gap value is lying between $\Delta = 20 - 30\,\mathrm{meV}$ for underdoped and optimally doped samples [120]. Even with inclusion of the copper oxide chain gap the spectroscopic results on Y123 in this work do not support an effective gap value of $> 30\,\mathrm{meV}$ because the total conductivity is already on the decrease at this energy.

In order to make sure that the introduction of the numerical scaling factor α has no poisoning effect on our extracted spin spectrum, the deconvolution of the Bi2212 spectrum by Gold's algorithm was performed for four different values of α. The results shown in Fig. 4.9 are comforting in the sense that the overall shape of the spin spectrum is unchanged. The only major difference lies in the magnitude of the zero-energy peak which is to be expected from a scalar multiplication, but since this peak is anyhow out of the bounds of physical contributions it does not harm the analysis. The oscillations in the obtained function $g^2\chi''$ for high energies are also not affected by a change of α. They are caused by the FFT of the elastic conductance, so their position and spacing only changes upon changing Δ. They are a side effect of the algorithm that might be reduced by clever filtering of the Fourier transformed elastic conductance that enters the algorithm.

4.5 Summary

The STM junction geometry was utilised to effectively couple to real bosons in inelastic tunnelling spectroscopy (IETS) on cuprate superconductors. The superconducting and normal state tunnelling spectra showed significant inelastic contributions for underdoped Bi2212 and optimally doped Y123. Since the effective Eliashberg function is inaccessible through the normal state in unconventional superconductors, it was obtained by deconvolution of the superconducting spectrum. In contrast to fully gapped unconventional superconductors like monolayer FeSe [27], the separation of elastic and inelastic tunnelling contributions proves to be more difficult in the nodal superconductors. Particularly, the k-dependence of Δ and the normal state DOS are important factors that were respected in a tight-binding model of the superconducting DOS based on ARPES measurements. By using the Gold algorithm for the deconvolution procedure, a strictly positive effective Eliashberg function was obtained. It takes the form of the expected integrated spin spectrum in the spin-fermion model with a resonance mode at $\omega_{res} > \Delta_0$. The resonance mode was found at $\omega_{res} \approx 70\,\mathrm{meV}$ in Bi2212 and at $\omega_{res} \approx 60\,\mathrm{meV}$ in Y123. Despite the loss of momentum information in STS, a description in momentum averaged electronic and bosonic spectral functions yields reasonable results because the tunnelling spectrum is dominated by the hot spots on the Fermi surface.

5 Yu-Shiba-Rusinov States and Long-lived Bosonic Excitations in Granular Aluminium

Granular aluminium (grAl) has recently established itself in parts of the superconducting qubit community as a suitable superinductor in Al/AlO$_x$ based superconducting circuits [29–31] and as a template for magnetic-field resilient qubit fabrication [32, 33]. Not only are excess quasiparticles identified as a limiting loss mechanism in these circuits, transport measurements in grAl near its MIT also hint at localized magnetic moments that might be responsible for the $1/f$ flux noise in Al/AlO$_x$ based junctions [34, 35]. STM provides a local probe of the quasiparticles on a microscopic scale and is sensitive to spin scattering, which should manifest as a pair of Yu-Shiba-Rusinov (YSR) states below T_c. In this chapter, the first low-temperature STM/STS results on grAl samples in two different resistivity regimes are presented and put into perspective by comparison with a polycrystalline pure aluminium film. The main focus in this work lies on the sub-gap YSR states that were found in the high resistivity grAl samples together with the inelastic excitations that are likely connected to phase modes of the superconducting order parameter. Lastly, an experimental procedure to systematically locate the spots, at which localized magnetic moments reside, is outlined and first results are shown.

Disclaimer: I, the author, want to emphasize that I was not personally involved in the measurements of the grAl samples but performed the measurements of the polycrystalline Al films (Chapter 5.2.1, 5.7). The measurements of the grAl samples were performed by the former group member Dr. Fang Yang and his at the time master student Tim Storbeck in the group of Prof. Wulfhekel at Karlsruhe Institute of Technology (KIT). I was involved in data analysis and presentation, as well as manuscript preparation for the publication that followed in Physical Review B. Parts of the results presented in this thesis chapter have been published by me as a co-author in [146]. Reused data is marked by the self-citation in the respective figure caption.

5.1 Relation between T_c and Oxygen Partial Pressure

The discovery of granular superconductors and their enhanced superconducting transition temperature, compared to the homogeneous bulk metal, has its roots in the group of Ben Abeles in Princeton [20, 147]. In the following years, grAl quickly became the centre of attention due to its controlled growth [148] and the large T_c enhancement to up to $3.15\,\mathrm{K}$ [149]. The term "granular superconductor" refers to a system in which superconducting grains are separated by insulating barriers, effectively building an array of intercoupled Josephson junctions (JJ). For aluminium, this is realized by evaporating aluminium onto a room temperature (or colder) substrate in an oxygen atmosphere. At oxygen partial pressures larger than roughly $1 \times 10^{-5}\,\mathrm{mbar}$ [148], the aluminium partially oxidizes in such a way that pure aluminium grains of a well-defined grain size a are separated by a non-stoichiometric AlO_x matrix. The normal state DC resistivity ρ_{dc} is directly proportional to the oxygen partial pressure and is the parameter of choice to compare samples grown in different setups. Up to the Mott resistivity $\rho_M \approx 400\,\mu\Omega\,\mathrm{cm}$, the mean size of the Al grains on room temperature substrates gradually shrinks to $a = (3 \pm 1)\,\mathrm{nm}$ and further remains constant for larger resistivities, yet with a smaller variance [150]. Close to the Mott resistivity, T_c peaks [151, 152] and reaches $2.17\,\mathrm{K}$ (for room temperature substrates) [153] before quickly decreasing for samples with $\rho_{dc} \gtrsim 3 \times 10^3\,\mu\Omega\,\mathrm{cm}$ due to a superconductor-to-insulator transition (SIT) [149]. This relation between normal state resistivity and T_c creates a superconducting dome, which is depicted in the phase diagram in Fig. 5.1. According to Ref. [153], the superconducting gap size Δ follows the general trend of T_c in its dependence on DC resistivity below ρ_M, but jumps near the Mott resistivity, which leads to a change of the BCS coupling strength Δ/T_c from ~ 1.78 to ~ 2.1 for the high resistivity regime [153]. The maximum gap size determined in their THz absorption experiments was $\Delta = (0.38 \pm 0.04)\,\mathrm{meV}$ for a sample with $\rho_{dc} = 1600\,\mu\Omega\,\mathrm{cm}$.

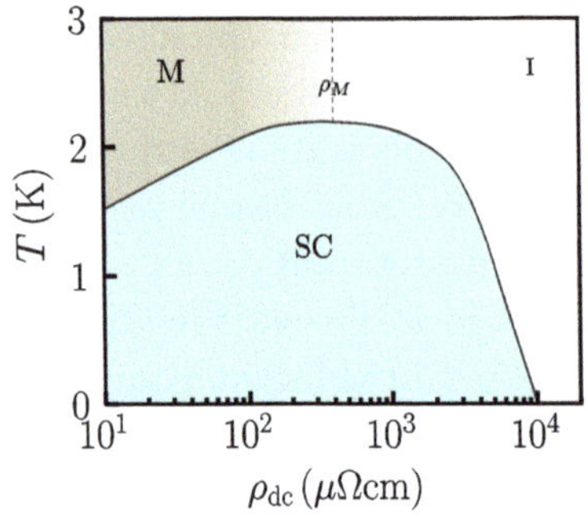

Figure 5.1: Superconducting Dome:
The superconducting transition temperature T_c of grAl (black line) gives rise to a superconducting dome in the phase diagram when plotted over the logarithmic normal state DC resistivity ρ_{dc}. It peaks at the Mott resistivity $\rho_M \approx 400\,\mu\Omega\,\mathrm{cm}$, at which the material features a MIT for $T > T_c$. Sketched according to data from Ref. [153].

Until today, no general consensus has been reached on the underlying mechanisms that govern the T_c enhancement, drive the SIT and allow superconductivity to survive for resistivities far above the Mott limit. A few widely accepted concepts shall, however, be briefly mentioned here. For further reading, the reader is referred to the correspondingly cited literature. Suggested mechanisms include

1. ..for the T_c enhancement:

   - *Size effects* of the discrete energy spectrum inside the grain lead to an effective BCS-like electron-electron interaction inversely proportional to the grain size and a transition to a strong coupling limit for small grain sizes [154].
   - *Shell effects* in the delocalized states of metallic nanosized clusters containing a specific number of atoms can lead to an increased pairing strength [155, 156].
   - *Phonon softening* due to the reduced dimension of the superconductor leads to a down shift of phonon DOS in energy and an increased Eliashberg electron-phonon coupling [157, 158].

   All these effects focus on an enhanced superconductivity of the metal grains. In the presence of the insulating barriers (AlO_x) they have to be combined with an intergrain coupling mechanism, e.g. a Josephson coupling like proposed in Ref. [154, 159, 160], in order to explain an increase in bulk T_c.

2. ..for the SIT:

   - *Quenching of superconductivity in individual grains* happens due to large amplitude fluctuations of the order parameter [161, 162] when the Josephson coupling between them is weak.
   - *Loss of macroscopic phase coherence* is caused by the systems increasing susceptibility to phase fluctuations of the order parameter (Experimental studies) [149, 153].

3. ..for the persistent superconductivity:

   - *Percolation* arguments can be made for the superconducting junctions not breaking down all at once but in a statistical manner leaving room for a few superconducting channels via well-coupled grains [160] (This can easily be discarded if a local STM probe of the LDOS shows a superconducting gap on each grain).

- The *Kondo effect* due to a singly occupied electronic level in the metal quantum dot circumvents the problem of the Coulomb blockade and restores metallicity via the Kondo resonance below T_K [163–165].
- *Strong-coupling effects* set in close to the MIT and can preserve superconductivity up to high resisitivities [165, 166]: Fermion pairs as small as the Fermion scattering length a_F keep a mean distance of k_F^{-1} to each other [167]. The transition to the dilute case $0 < k_F a_F \ll 1$, in which the mean number of Cooper pairs inside a coherence volume is of order unity ($k_F \xi \lesssim 1$) [168], is therefore connected with a BCS-BEC crossover [167, 169].

5.2 Microscopic Grain Coupling and the Superconducting Gap

The aluminium and granular aluminium films were grown on Nb doped (0.7 weight %) $SrTiO_3$ (Nb:STO) single crystals of (001) orientation. The substrate provides a conducting, flat and nonreactive surface. Aluminium was deposited by electron beam evaporation in a commercial PLASSYS MEB550 at a rate of $1\,\mathrm{nms}^{-1}$ up to the desired film thickness and the grAl films were grown in an oxygen atmosphere. The growth parameters and estimated sample resistivities are summarized in Tab. 5.1. Subsequently, the samples were immediately transferred to the STM setup via a vacuum suitcase to keep surface contamination to a minimum.

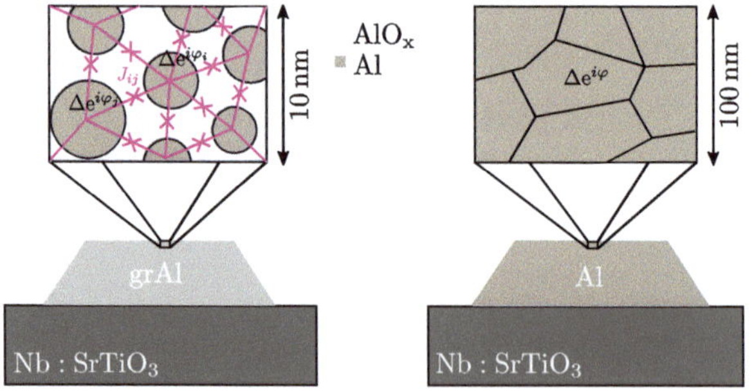

Figure 5.2: Sample Structure: Sample constituents and schematic structures: Left: The grAl samples feature nano-sized aluminium grains embedded in an amorphous AlO_x matrix. The coupling of individual superconducting Al grains essentially leads to a large array of Josephson junctions. Right: The polycrystalline Al film features larger crystallites and macroscopic phase coherence of the superconducting order parameter.

Table 5.1: (Granular) Aluminium Samples: Investigated samples including their estimated resistivity ρ_{dc} at oxygen partial pressure p_{O_2} and film thickness d following established growth recipes further detailed in Ref. [29].

Sample tag	Material	$\rho_{dc}(\mu\Omega\,cm)$	p_{O_2} $(10^{-5}\,mbar)$	$d\,(nm)$
A	Al	< 10	0	30
G1	grAl	300	2.8	50
G2	grAl	2000	5.6	50

Fig. 5.2 schematically summarizes the most important structural properties that are expected from the grAl and Al films. Based on early electron microscopy studies [148, 150], the grAl film should consist of nano-sized Al grains separated by an amorphous AlO_x matrix. The thickness of the oxide, which is fine-tuned by the oxygen partial pressure during growth, eventually causes the SIT. As long as Cooper pairs can tunnel between two neighbouring Al grains i and j with order parameter $\Delta e^{i\varphi_{i,j}}$, the material is superconducting and acts as a large array of Josephson junctions with individual Josephson couplings J_{ij}. Above a critical thickness the grains uncouple and the material becomes insulating. The polycrystalline Al film should consist of larger crystallites due to the high diffusion mobility of Al at room temperature [146]. There, the phase of the superconducting order parameter is expected to be universal throughout the sample.

5.2.1 Pure Al Film

The measurement of sample A, the pure Al film, served as a control experiment. Fig. 5.3(a) shows the topographic image of the film that features crystallites of about 50 nm lateral size. The large-range $I(U)$ curve in Fig. 5.3(b) confirms the metallic behaviour for large bias voltages. Fig. 5.3(c) shows the low bias dI/dU spectrum at $T = 30\,mK$ which reveals a superconducting gap of $\Delta = (191 \pm 1)\,\mu eV$. The gap size was determined by a fit to the BCS DOS. This is close to the value of $200\,\mu eV$ reported for a $42\,nm$ thick Al film [170]. In all structural and electronic aspects, the sample behaves as expected.

5.2.2 Oxygen Poor GrAl Film

Next, the oxygen poor grAl sample G1 with an estimated resistivity of $\rho_{dc} = 300\,\mu\Omega\,cm$ was investigated. As can be seen in the topographic image in Fig. 5.4, a granular structure of round Al grains with sizes ranging from $5 - 10\,nm$ in diameter formed.

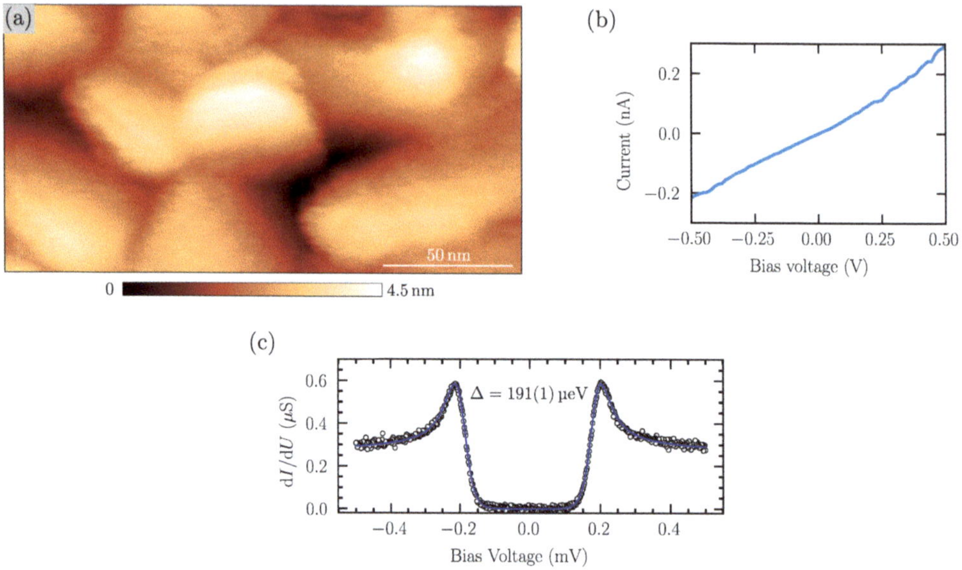

Figure 5.3: Pure Al Sample: (a) Topographic image of the pure Al film ($U = -1\,\mathrm{V}$, $I = 100\,\mathrm{pA}$) showing crystallites of about $50\,\mathrm{nm}$ lateral size. (b) Typical large range $I(U)$ curve displaying ohmic behaviour of the sample. (c) $\mathrm{d}I/\mathrm{d}U$ spectrum at $T = 30\,\mathrm{mK}$ showing the superconducting gap of $\Delta = (191 \pm 1)\,\mathrm{\mu eV}$. The gap size was determined from the fit of a BCS gap (blue line). Data previously published in [146].

For objects this small and pronounced, the shape of the tip plays an important role in the topographic image [171]. In fact, the resultant topography in the constant current mode reflects a convolution of tip surface and sample surface. The tip cannot be etched to arbitrary sharpness and tip apex radii of a few nm are not unusual [112, 172]. Taking this into account, the observation agrees well with previously reported grain sizes and spread for similarly grown grAl films [150]. A comparison between typical $I(U)$ curves measured on different grains of G1 (Fig. 5.4(b)) and the $I(U)$ characteristic of the pure Al film A (Fig. 5.3(b)) confirms that sample G1 is still in the metallic regime.

Fig. 5.4(c) shows the $\mathrm{d}I/\mathrm{d}U$ spectrum at $T = 30\,\mathrm{mK}$ with a superconducting gap of $\Delta = (298 \pm 1)\,\mathrm{\mu eV}$, which is significantly larger than for the pure Al film. The gap size was determined by a BCS fit (blue line). The gap size was practically identical when probed on different grains (see Fig. 5.4(d)). For similar grAl films, the authors of Ref. [173] derived $\Delta = 344\,\mathrm{\mu eV}$ from T_c. THz spectroscopy experiments yielded $\Delta = (336 \pm 8)\,\mathrm{\mu eV}$ [153]. The uniform gap size on the surface for different grains could be explained by either a homogeneous gap enhancement mechanism which is independent of the grain size or a strongly coupled network in which the Δ of different grains are aligned. Either way, grains on the surface would exhibit

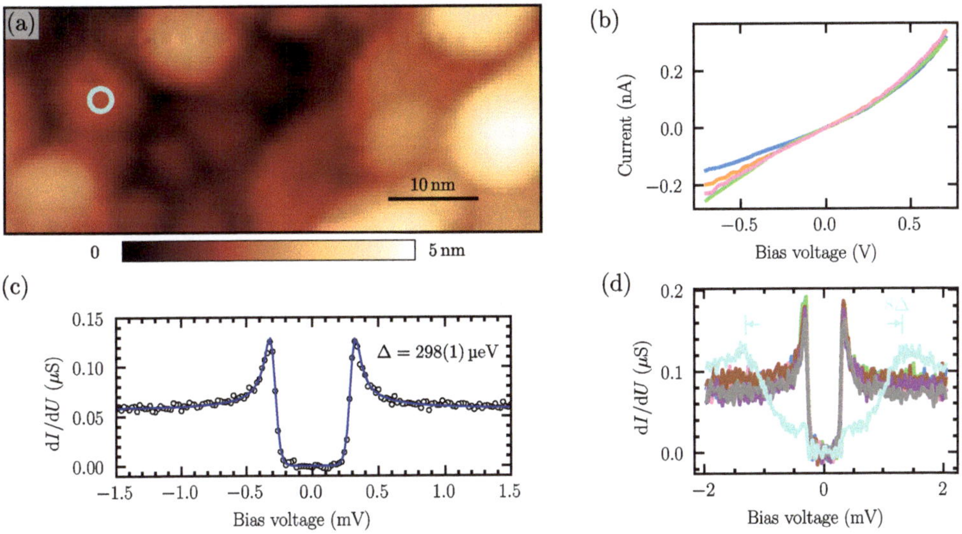

Figure 5.4: Oxygen Poor Sample: (a) Topographic image of the grAl film ($U = 60\,\mathrm{mV}$, $I = 240\,\mathrm{pA}$) with $\rho_{\mathrm{dc}} \approx 300\,\mu\Omega\,\mathrm{cm}$ showing grains of $5 - 10\,\mathrm{nm}$ lateral size. (b) Typical large range $I(U)$ curves displaying metallic behaviour of the sample. (c) $\mathrm{d}I/\mathrm{d}U$ spectrum at $T = 30\,\mathrm{mK}$ showing the superconducting gap of $\Delta = (298 \pm 1)\,\mu\mathrm{eV}$. The gap size was determined from the fit of a BCS gap (blue line). (d) $\mathrm{d}I/\mathrm{d}U$ spectra on different grains reveal a practically universal gap except for the grain marked in cyan in (a), which exhibits an additional pseudogap of roughly $8\,\Delta$. Data previously published in [146].

the same gap as in the bulk making the STS measurements here comparable to bulk measurements. In Ref. [173], where Δ was derived from T_c measurements, an overestimation can be explained by the overall higher T_c compared to Ref. [153], which indicates a significant difference in the systematic approach. It should be noted that THz spectroscopy represents a real bulk measurement of Δ, whereas in STS the sample LDOS at the tip position is measured. Other factors, that come into play, are: 1. the overall lower oxygen concentration near the surface for the UHV samples used in this STM study compared to the ambient samples from different studies, 2. the overall higher oxygen concentration at the surface compared to the bulk due to the growth procedure, 3. the fact that only a small area on the surface was probed by STM and 4. the limited comparability of samples produced by different recipes and in slightly different environments. Considering all these factors, the agreement between the gap size determined here and bulk measurements is surprisingly good.

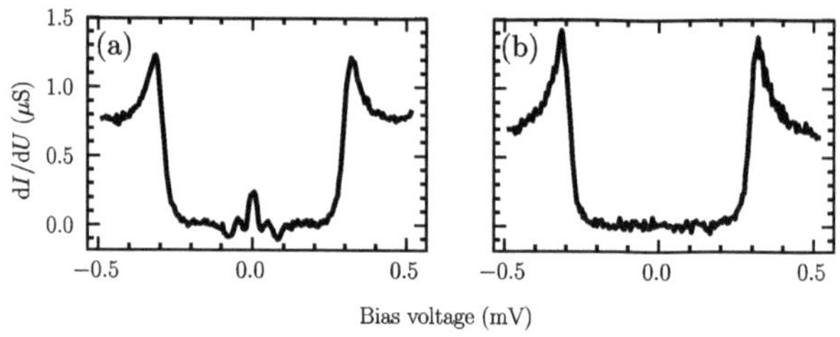

Figure 5.5: Josephson Tunnelling: $\mathrm{d}I/\mathrm{d}U$ spectrum under extreme tunnelling condition (high tunnelling rate) on the insulating grain (a) from Fig. 5.4(a) and a more metallic grain (b). Tip stabilization condition: $U = 500\,\mu\mathrm{V}$, $I = 200\,\mathrm{pA}$. Data previously published in [146].

On the grain marked in cyan in Fig. 5.4(a), the low energy spectrum shows an additional pseudogap[1] of roughly $8\,\Delta$, as shown as by the cyan line in Fig. 5.4(d). As we will see, this behaviour is rather atypical for the oxygen poor sample but not uncommon for the oxygen rich sample. The LDOS for $eU > \Delta$ is heavily reduced in comparison to the other grains in this sample. For extreme tunnelling conditions on this insulating grain, i.e. a conductance at the scale of μS and a tunnelling current approaching the critical current, the $\mathrm{d}I/\mathrm{d}U$ spectrum shown in Fig. 5.5 (a) was obtained. It features a sharp peak at zero bias, flanked symmetrically by in-gap peaks and a dip to negative differential conductance. The spectrum for a metallic grain at the same tunnelling conditions does not show these features (Fig. 5.5(b)). Under extreme tunnelling conditions, the voltage drop over the vacuum barrier seems to become comparable to the voltage drop over the sample. As a consequence, the measured $\mathrm{d}I/\mathrm{d}U$ spectrum also contains information on the electronic transport inside the sample. The zero-bias peak is reminiscent of the Josephson DC current in an SIS junction, which can be understood as follows: A Cooper pair in one superconductor is destroyed through an Andreev reflection at the insulator interface and continues as an evanescent wave in the insulator. If the evanescent wave overlaps with the one coming from the opposite superconductor, then a Cooper pair can be created in the opposite superconductor by an additional Andreev reflection of the electron-hole pair. Hence, a dissipationless DC current flows through the superconductor. The oscillations and negative differential conductance at $e|U| \gtrsim 0$ are consistent with previous STS measurements of the Josephson DC current and explained within the $P(E)$ theory [106, 175]. Direct lead-to-lead elastic cotunnelling and non-local Andreev reflections [176, 177], which cause very similar $\mathrm{d}I/\mathrm{d}U$ features [178], are

[1] Although the term "pseudogap" is nowadays often associated with high-temperature superconductivity, it is here meant in the original sense popularized by Mott, i.e. a depletion of electronic states near the Fermi energy due to increased localization of electronic wave functions near an MIT [174].

not expected because the superconducting coherence length in grAl is reduced to about 10 nm [179], so below the film thickness. If the additional features near zero bias voltage stem from intergrain tunnelling processes, then the $\mathrm{d}I/\mathrm{d}U$ spectrum should also exhibit coherence peaks at 2Δ, due to the doubled energy needed to transfer a quasiparticle from one superconductor to another superconductor with the energy gap Δ. Unfortunately, no data in this bias range was recorded. It leaves open the question, whether the acquired $\mathrm{d}I/\mathrm{d}U$ spectrum is the superposition of the LDOS on the grain and intergrain tunnelling characteristics in the bulk or a different phenomenon.

5.2.3 Oxygen Rich GrAl Film

In Fig. 5.6, the general behaviour of the oxygen rich film with $\rho_{\mathrm{dc}} = 2000\,\mu\Omega\,\mathrm{cm}$ is summarized. The topographic image in Fig. 5.6(a) confirms that its grain sizes are similar to the oxygen poor film's, an observation well in agreement with previous electron microscopy studies [150]. The much higher resistance of this sample is, however, reflected in the fact that $I(U)$ curves (Fig. 5.6(b)) measured on individual grains, marked in Fig. 5.6(a), exhibit insulating behaviour, so a region of almost zero current for voltages lower than a threshold of roughly 70 to 300 mV (depending on the grain and bias sign). Above the threshold voltage, a differential conductance ($I(U)$ slope) comparable to the oxygen poor sample is obtained. The sudden onset of electronic transport above the threshold voltage signals that the grains are in the Coulomb blockade regime. At smaller tunnelling rates (lower tunnelling conductance) the $I(U)$ spectrum evolves to a Coulomb staircase, which is reflected in the $\mathrm{d}I/\mathrm{d}U$ spectrum in 5.6(c) as resonant tunnelling peaks for $e|U| > E_c$, where E_c is the charging energy [180].

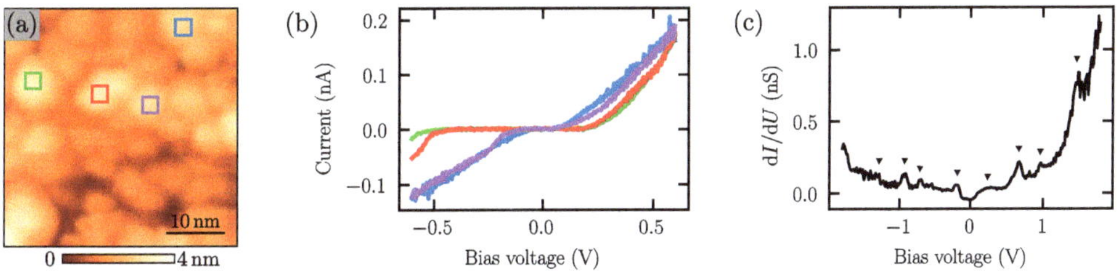

Figure 5.6: Oxygen Rich Sample: (a) Topographic image of the grAl film ($U = 400\,\mathrm{mV}$, $I = 180\,\mathrm{pA}$) with $\rho_{\mathrm{dc}} \approx 2 \times 10^3\,\mu\Omega\,\mathrm{cm}$ showing grains of $2 - 5\,\mathrm{nm}$ lateral size. (b) Typical large range $I(U)$ curves obtained at the grains indicated by the same colors in (a). The $I(U)$ curves display insulating behaviour of the sample with varying gaps. (c) $\mathrm{d}I/\mathrm{d}U$ spectrum of a particularly isolated grain in the Coulomb blockade regime showing charging peaks at the energetic positions indicated by triangular markers. Data previously published in [146].

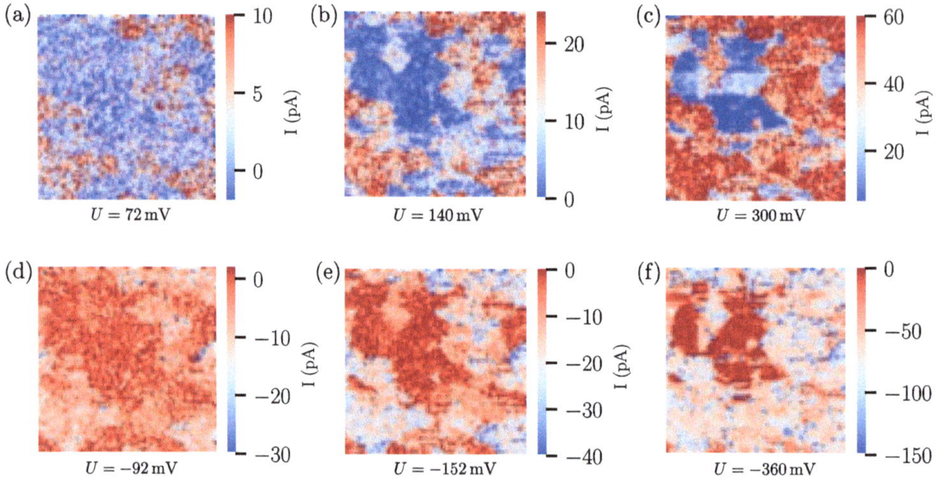

Figure 5.7: Grain Coupling: (a-f) Constant height maps of the area shown in Fig. 5.6(a) at the indicated bias voltages U display the local tunnelling current I. Multiple neighbouring grains form electrical clusters that charge and discharge collectively. Data previously published in [146].

The energy of these peaks corresponds to first, second, etc. charging state of the single grain or cluster [181]. Using

$$E_c = \frac{e^2}{2C}, \tag{5.1}$$

the first charging state at $U \approx \pm 0.2\,\mathrm{V}$ in Fig. 5.6(c) corresponds to a capacitance of $C \approx 0.4\,\mathrm{aF}$. The threshold voltages or charging energies ranging from ± 70 to $\pm 300\,\mathrm{meV}$ thus correspond to capacities between 0.3 and $1.1\,\mathrm{aF}$. Approximating the capacitor as a sphere, of which one half is embedded in a dielectric of permittivity $\varepsilon_r = 9$ and thickness $0.5\,\mathrm{nm}$, the smallest capacitance corresponds to a charged object of approximately $2\,\mathrm{nm}$ in diameter (see Sec. 5.5 for the model). Apparently, the smallest chargeable object is a single grain, but larger clusters of grains can also form electrical islands that are simultaneously charged and discharged. This conclusion is supported by the tunnelling current maps shown in Fig. 5.7. The maps were recorded in the constant-height-mode on the area from Fig. 5.6(a). An abrupt increase in tunnelling current for increasing bias voltage is seen at once for several neighbouring grains, indicating that several grains indeed form strongly coupled electrical clusters. Recent measurements of a transmon qubit made from thin film grAl also find evidence for a strong electrical coupling of about 10 grains to one effective Josephson junction [32].

The detailed discussion of the low energy spectrum and especially the superconducting gap in sample G2 is shifted to the sections 5.3 and 5.4, which focus on the in-gap and out-gap structure of the DOS respectively. It should only be aforementioned here, that the gap size for this sample is also enhanced by the granularity and with $\Delta = (312 \pm 1)\,\mu\text{eV}$ also slightly larger than for sample G1, which is in agreement with optical spectroscopy measurements [153] on samples of similar resistivity.

5.3 Spectroscopic Signs of YSR states

Fig. 5.8(a) shows a typical high-resolution spectrum of the superconducting gap in sample G2. A BCS fit (blue line) yields a superconducting gap of $\Delta = (312 \pm 1)\,\mu\text{eV}$. Remarkable in this spectrum are the two in-gap states at energetically symmetric positions with respect to the Fermi energy, which are indicated by black triangles. These in-gap states are found for eleven out of 23 individually probed grains. The intensity of these peaks is low compared to the differential conductance outside the gap. Their energetic position is very close to Δ and thus rules out that they are resonances due to Andreev bound states (cf. Fig. 5.5). The observation of the Kondo effect in grAl films for temperatures larger than T_c [35, 163] indicates that the resonances seen here are spectroscopic signs of Yu-Shiba-Rusinov (YSR) states. In simplest terms they can be thought of as the analogon to the Kondo effect for a superconductor. They are caused by the exchange coupling between conduction electrons and unpaired localized electrons or magnetic moments in the presence of

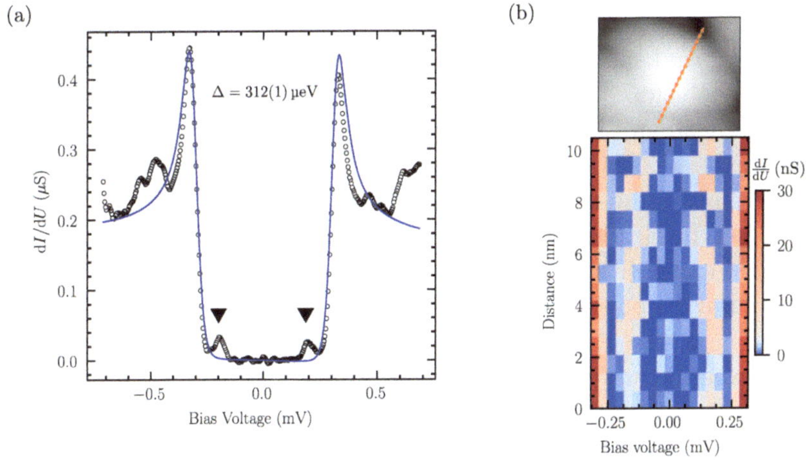

Figure 5.8: YSR States: (a) dI/dU spectrum on an individual grain of the oxygen rich sample. Triangular markers indicate the positions of YSR states inside the gap. (b) Heatmap of the dI/dU spectrum as a function of lateral tip position across a single grain. The line profile is indicated in orange on top of the grey-scale topographic image. YSR states move symmetrically in energy as a function of the tip-grain distance. Data previously published in [146].

a superconducting gap. In this scenario, a Cooper pair can be broken for energies smaller than Δ if the resulting pair of quasiparticles participates in the screening of the localized spins and thus compensates the cost of kinetic energy by a reduction of the mean potential energy.

As pointed out by the authors of Ref. [35], two scenarios for the origin of unpaired magnetic moments have to be distinguished: They can either reside on the Al grains or in the insulating oxide. In the former scenario, a sufficient decoupling of individual grains in the Coulomb blockade regime can result in an odd number of electrons on individual grains in the equilibrium state. These so-called Kubo spins are suggested as the localized moments responsible for the Kondo effect at $T > T_c$ by the authors of Ref. [182]. It still remains unclear, how exactly these localized and unpaired electrons would survive the onset of superconductivity, where electrons in the nanosized Al grains pair up to Cooper pairs. The quenching of superconductivity in a few grains by a combination of size effects and decoupling from the rest of the network at first glance contradicts the observed large T_c and enhanced electron-phonon coupling [153] for $\rho > \rho_M$. In the latter scenario, the nonstoichiometric composition of AlO_x could lead to trapped, unpaired electrons in dangling bonds at surfaces or interfaces. The existence of these magnetic impurities would not be altered by the onset of superconductivity and their abundance would scale with the thickness of the oxide, so the resistivity of the samples.

In order to figure out, which scenario is present, the dependence of the YSR states on an electric field was measured. This was realized by moving the tip laterally across a single grain and recording the full $\mathrm{d}I/\mathrm{d}U$ spectrum in equidistant steps. The line profile and spectroscopic positions with respect to the grain are shown in Fig. 5.8(b) in the top panel. In the lower panel the $\mathrm{d}I/\mathrm{d}U$ spectrum is shown in the form of a heatmap. While the superconducting gap edge (dark red) remains fixed, the YSR states (apricot colour) vary in energy as a function of tip position. If the unpaired spins resided in electrically decoupled grains with an odd number of electrons, then the electric field of the biased tip-sample junction would shift the chemical potential of the grain. As a result, the energetic position of the gap edge would move but the position of the YSR states with respect to Δ would stay fixed. This contradicts the results shown here. Instead, unpaired spins in the oxide could explain the experimental results. The voltage drop over the insulating oxide barrier between the grains is affected by the electric field at the surface. With a variation of the voltage drop across the insulator, the energy of the unpaired spins is changed and ultimately their exchange coupling to the conduction electrons is modified. As demonstrated by the authors of Ref. [183], this effect leads to an energy shift of the YSR states with respect to Δ as a continuous function of the exchange coupling.

5.4 Spectroscopic Signs of Long-lived Plasma or Higgs modes

Turning the attention to the out-gap structure of the differential conductance displayed in Fig. 5.9 (black line), which was recorded on sample G2, a pseudogap like in Fig. 5.4(d) of $> 8\Delta$ is retrieved[2]. Only now, a previously absent or invisible series of peaks inside this pseudogap is apparent. The energetic positions and width of these peaks are too small to be caused by charging effects, like in Fig. 5.6(c). The sharpness of the peaks is not in agreement with simple pair-breaking by inelastic tunnelling, which would make these excitations strongly damped because the electron pairs quickly recombine. Instead, their shape resembles the coherence peaks. Out-gap copies of the coherence peaks are indicators of inelastic tunnelling processes involving a rather long-lived bosonic mode: The incident tunnelling electron excites a boson of defined energy and the final scattered electron with $E \gtrsim \Delta$ ends up in the quasiparticle resonance of the BCS single particle DOS. Due to the high DOS of the final state electron, this process has a large weight and prominent copies of the coherence peaks, shifted from their original position by the boson energy, are the consequence. The first copy corresponds to the creation of one boson with energy ω, the second to two bosons, etc.

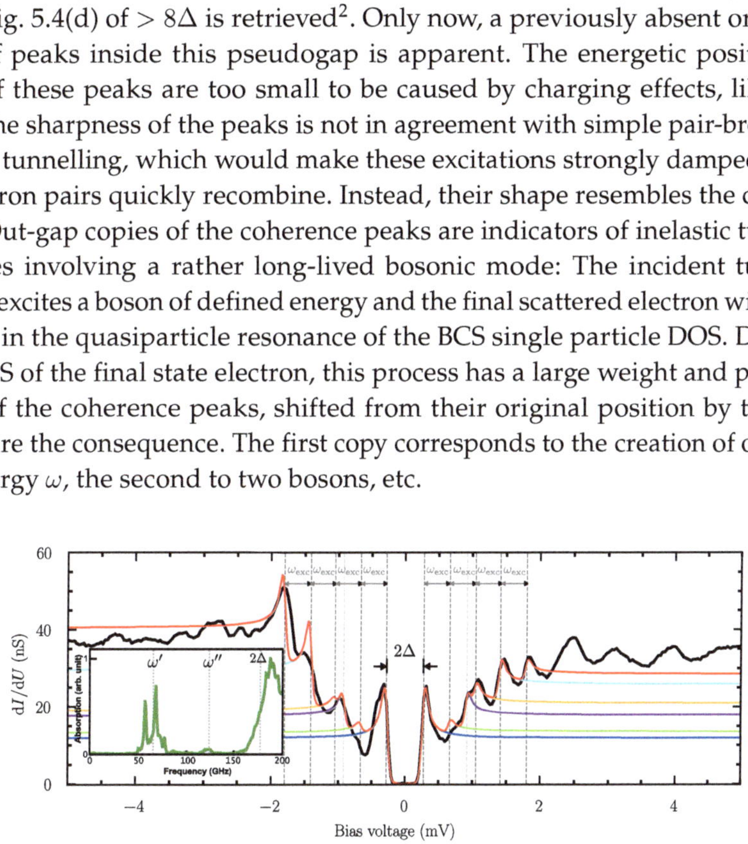

Figure 5.9: Inelastic Excitations: High resolution dI/dU spectrum (black line) featuring a superconducting gap and several inelastic excitations which appear as copies of the coherence peak for $eU > \Delta$. The excitation ladder can be modeled by five energetically symmetric inelastic excitations and fit to the spectrum (red line). The modeled spectrum for zero, one, two, etc. excitations is individually displayed in blue, green, violet, and so on. Except for the excitation shown in violet, all excitation energies are integer multiples of ω_{exc}. The energies $\Delta + n \cdot \omega_{\text{exc}}$ are indicated by dark grey dashed lines. The inset reproduces THz spectroscopy measurements of samples with $\rho \approx 2000\,\mu\Omega\,\text{cm}$ from Ref. [153]. Data previously published in [146].

[2] The discussion of this pseudogap is postponed to the end of this section.

Table 5.2: Fit Parameters for Bosonic Excitations: Fitted parameters of Eq. (5.2) used to model the spectrum in red in Fig. 5.9. n is the number of the bosonic excitation, ω_n its energy divided by $\Delta = (277 \pm 4)$ µeV and a_n^+ and a_n^- are the intensities for positive and negative bias range.

n	ω_n (Δ)	a_n^+	a_n^-
1	(1.40)*	(0.142)	(0.118)
2	2.31	0.369	0.377
3	2.78	0.259	0.111
4	4.09	0.418	0.899
5	5.50	0.223	0.933

*This mode was manually set to this energy from comparison to the experimental curve. It is found at a too high energy ($\sim 1.8\Delta$) by the fit because the BCS fit exceeds the experimental curve.

In the simplest form, the resultant $\mathrm{d}I/\mathrm{d}U$ spectrum of this excitation ladder can be modeled by fitting a BCS DOS to the gap for $E \lesssim 2\Delta$, which should constitute the elastic tunnelling signal σ_{el}, and then adding this functional form shifted by the respective energy of the inelastic excitation with a variable intensity a_n. For N inelastic conduction channels involving bosons with energy ω_n, the fit function shown as a red line in Fig. 5.9 reads

$$\sigma_{\mathrm{tot}}(U) = \sum_{n=0}^{N} a_n \sigma_{\mathrm{el}}(U - \omega_n/e) \tag{5.2}$$

with $a_0 = 1$ and $\omega_0 = 0$ for the purely elastic part. This approach corresponds to a boson spectrum consisting of infinitely sharp modes, so single δ-distributions at ω_n. The experimental curve (black line) was fitted to this model separately for the positive and negative bias range. For $U > 0$, the bosonic energies ω_n and intensities a_n were left as adjustable fit parameters. For the negative bias range, the bosonic energies, which were found for the positive bias range, were enforced, but the intensities were still left adjustable. This choice was totally ambiguous and has also been performed in reverse yielding very similar bosonic energies. As obvious from the measured spectrum, the peaks appear at voltages symmetric with respect to zero bias. Their intensities, however, are different for positive and negative bias voltage. The shape of the spectrum is nicely followed by the fitted function (red line) taking into account $N = 5$ boson excitations without additional lifetime broadening. The curves in blue, green, violet, yellow and light blue show the fit up to the n'th excitation. The parameters of the fit function are summarized in Table 5.2.

Apart from the second excitation, shown as a violet line, all excitation energies are roughly equidistantly spaced in the spectrum. This is indicated by the dark grey dashed lines in Fig. 5.9. These energies are (within the margin of error) multiples of $\omega_{\mathrm{exc}} \approx 1.4\Delta$. A dissipative mode of the superconductor at this energy was also found in THz absorption experiments on films of similar sheet resistance, as shown in the inset of Fig. 5.9. The above-mentioned mode is labeled ω''. While the modes at $\omega' < \Delta$ were assigned to two-dimensional plasma phase modes [153, 184], the origin of the resonance at $\Delta < \omega'' < 2\Delta$ remains unclear. A phononic origin of the mode at energy $\omega_{\mathrm{exc}} \sim \omega''$ can be ruled out as van Hove singularities in the phonon DOS of Al or AlO_x all lie above 15 meV [185]. The same holds for the vibrational modes of chemisorbed oxygen on the aluminium surface [186]. In 3D bulk superconductors the NG mode is usually gapped up to the plasma frequency ω_p (so to much higher energies than Δ). The drastic reduction to sub-gap energies in grAl samples with high sheet resistance, is ascribed to the granularity [184, 187] and the quantisation of two-dimensional plasmon modes [153] that have maximum DOS at the effective plasma frequency[3]. While bulk plasmonic excitations are expected above ω_p, the sharp peaks in energy seen in this work speak more for a non-dispersive mode with a long lifetime.

Recent theoretical studies predict, that the mass of the Higgs mode can be decreased below 2Δ in disordered superconductors [189], making it another plausible candidate. In some of these disordered superconductors near an SIT, Higgs modes with long lifetimes and a renormalized mass just below 2Δ were reported [64, 190]. The mechanism, that was predicted in numerical calculations is, however, still missing confirmation by an analytic theory [191]. It has been shown that in the strong coupling limit, which we expect to be in beyond the Mott resistivity, amplitude fluctuations become equally important as phase fluctuations [98]. The energy scale at which this happens is $T \sim \omega_{\mathrm{exc}}$, which is for the excitation energy determined here, roughly twice the critical temperature. This might be an indicator for the existence of an additional temperature scale $T_{\mathrm{pair}} > T_c$, in which pairing is already present locally but not in a coherent manner due to large fluctuations. Similar to the case of the cuprates, this would induce a pseudogap above T_c. Such a pseudogap was already proposed as an explanation for the anomalous behaviour of Δ and the real part of the dynamical conductivity in THz spectroscopy of oxygen-rich grAl samples [149]. Further discussion on the origin of this long-lived mode is postponed to Sec. 5.5 and 5.6, in which our tunnelling data is compared to the predictions within a concrete model for plasmon modes.

In a more general approach, the $\mathrm{d}I/\mathrm{d}U$ spectrum from Fig. 5.9 was also separated into an elastic and inelastic part and the bosonic spectral density function $\alpha^2F(\omega)$ was determined analagously to how the spin spectrum was obtained for the cuprate superconductors in Chap. 4. This approach has the advantage that the functional

[3] In 2D, plasmons are not gapped [188].

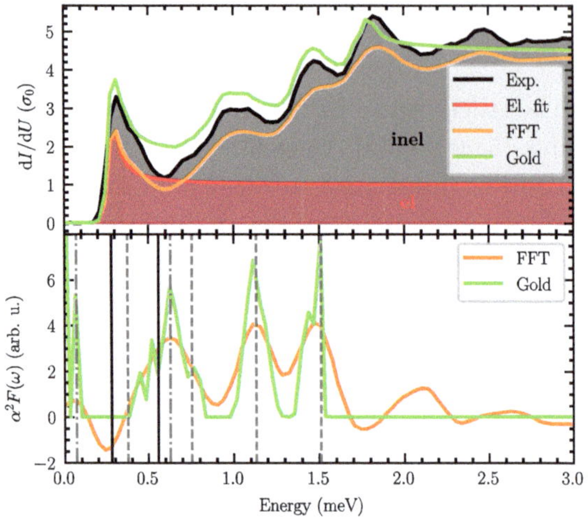

Figure 5.10: Boson Spectral Density: Top panel: A separation of the experimentally determined $\mathrm{d}I/\mathrm{d}U$ spectrum (black line) into an elastic part (red) and an inelastic part (grey) allows for a deconvolution by means of FFT and Gold algorithm. The forward calculated conductance spectrum using the bosonic spectral density function obtained from FFT/Gold algorithm is shown in orange/green. Bottom panel: Bosonic spectral density function obtained via FFT/Gold algorithm (orange/green). Black lines indicate Δ and 2Δ. Dashed lines indicate the energies $n \cdot \omega_{\text{exc}}$. Dashed-dotted lines indicate additional resonances at ω^* and $2\Delta + \omega^*$.

form of the bosonic spectrum is not limited to a series of δ-distributions and the number of modes is a result of the deconvolution, not postulated a priori. First, the spectrum from Fig. 5.9 was symmetrized around zero bias in order to enhance the sensitivity for even components of the spectrum, i.e. the particle-hole symmetric parts including the peaks of inelastic excitations. For detailed information on how the data is prepared and the deconvolutions are performed, the reader is referred to Chap. 4 and Appendix A. The symmetrized spectrum is shown in Fig. 5.10 in the top panel (black line) and a BCS fit to the gap (downscaled to an intensity which is always lower than the measured curve) represents the elastic tunnelling conductance (red line). Subtracting this elastic conductance from the experimental curve yields the inelastic tunnelling conductance (grey area) which is a convolution of the bosonic spectral density and the elastic conductance. The determination of the bosonic spectral density $\alpha^2F(\omega)$ was performed via direct fast Fourier transform and the Gold algorithm and is displayed in the bottom panel in orange and green respectively. The respective results of the forward convolution using this bosonic function yields the total conductance in orange/green in the top panel.

Like in Chap. 4, the bosonic spectrum acquired through the FFT deconvolution (orange) contains non-zero values for $E < 0$, that are cut off in the linear forward convolution. This leads to a calculated conductance below the experimental curve. Also, once again, unphysical negative values are obtained in several energy regions including a region around Δ. Despite these shortcomings, the obtained bosonic spectrum in orange already reveals three broad but prominent modes centred at $0.63\,\mathrm{meV}$, $1.12\,\mathrm{meV}$ and $1.51\,\mathrm{meV}$ and the experimental $\mathrm{d}I/\mathrm{d}U$ spectrum is qualitatively well reproduced with it. The bosonic spectrum obtained via Gold algorithm (green) is again limited to strictly positive values in agreement with the physical constraint. The three above-mentioned modes are also found at the same energies, yet with smaller width. Around the peak at $0.63\,\mathrm{meV}$, additional side peaks at $0.455\,\mathrm{meV}$, $0.525\,\mathrm{meV}$ and $0.77\,\mathrm{meV}$ are found. One low energy mode at $0.07\,\mathrm{meV}$ is found at a position where the FFT curve also exhibits a small peak. Although slightly above the experimental conductance curve, the forward convolved $\mathrm{d}I/\mathrm{d}U$ curve nicely reproduces the observed peaks in position and shape up to $2\,\mathrm{meV}$. A closer look at the mode energies in relation to the gap size Δ reveals that the peak energies $\omega_2 = 0.77\,\mathrm{meV}$, $\omega_3 = 1.12\,\mathrm{meV}$ and $\omega_4 = 1.51\,\mathrm{meV}$ correspond to approximately $n = 2\,,3\,,4$ times $\omega_{\mathrm{exc}} = \omega_1 = 0.37\,\mathrm{meV} = 1.35\,\Delta$, which marks the onset of spectral density between Δ and 2Δ. These energies are therefore connected to n boson excitations with energy ω_{exc} and rising in amplitude with the number n. They are marked by grey dashed lines in the bottom panel of Fig. 5.10. High spectral density at 2Δ is expected to result from efficient pair breaking by inelastic tunnelling processes. The peak amplitude is found at $2\Delta + \omega^* = 0.525\,\mathrm{meV}$, with $\omega^* = 0.07\,\mathrm{meV}$ being the sub-gap mode. Since the experiment only yields boson mode energies with no ties to their origin, it remains unclear whether these two modes are connected with each other or not. They are marked as grey dot-dashed lines in the bottom panel of Fig. 5.10. They might also be falsely appointed to the boson spectral density and are really part of the elastic conductivity, i.e. the sample LDOS.

In order to gain more confidence in the presented conclusions on the boson energies and prevent an over-analysis of the data, an independent measurement at a different tip position with different stabilization parameters and bias step size was repeated. The results are shown in Appendix B. The analogous analysis, especially the deconvolution, independently confirms the bosonic mode at $\omega_{\mathrm{exc}} \sim 1.35\,\Delta$ and even the first excitation is clearly visible there. In total, five prominent excitations are seen, from which the first three fall on multiple energies of ω_{exc} and the last two are slightly shifted to higher energies. Like in the previous spectrum, the weight of the multi-boson excitation seems to grow up to $n \sim 3 - 4$ before decreasing again for higher n.

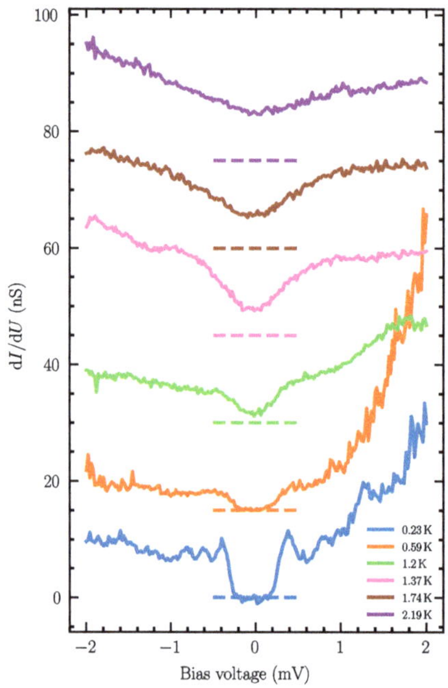

Figure 5.11: Temperature Dependence of Low Energy Spectrum: dI/dU spectra at rising temperatures from well below T_c (blue) to above T_c (purple). Individual spectra are offset with respect to each other by 15 nS. Dashed lines indicate the corresponding base line of zero conductance. Data previously published in [146].

At last, the larger pseudogap, which was found with a size of $\sim 8\Delta$ (without fine structure) in a single more insulating grain of sample G1 and with a size of $\sim 14\Delta$ (with a series of equidistant peaks) in the high resistivity sample G2 seems to be shaped by bosonic excitations of the condensate. Whether they are phase or amplitude fluctuations of the superconducting order parameter cannot be concluded from the available data. Such a depletion of states near the Fermi level would happen on a larger energy scale if initiated by the Coulomb blockade in the system. In THz spectroscopy measurements of high-resisitivity grAl samples, the anomalous behaviour of gap size and the real part of the dynamical conductivity has been attributed to a pseudogap above T_c as a consequence of a phase-fluctuation driven SIT [149]. This is at least in agreement with the results shown here in the sense that this pseudogap only arises in the high-resistivity regime and seems to grow with increasing resistivity. As order parameter fluctuations and long-range order are competing, a coexistance of such a pseudogap and the superconducting gap below T_c is not too surprising and can be understood as a foretaste of what is to happen at larger grain decoupling at the SIT. How amplitude fluctuations in the small grain regime can destroy superconductivity has been outlined by Guy Deutscher in the

70's [161] and more rigorously explained by Chubukov [98]. Here, measurements above T_c could be helpful to understand the connection between this pseudogap and superconductivity.

In order to see if inelastic excitations still play a large role above T_c, tunnelling spectra were measured as a function of temperature between 230 mK and 2.19 K. The results are shown in Fig. 5.11 and are offset to each other by 15 nS. The respective base line of zero conductance is indicated by a dashed line for each temperature. Up to 590 mK, the measured grain is superconducting, the gap is fully developed. Above 1.2 K, the measured in-gap conductance does not reach zero anymore due to temperature broadening of the reduced gap. This, however, does not mean that bulk T_c has already been reached. Between 1.37 K and 2.19 K the spectra exhibit a larger bias range of decreased conductivity that gradually broadens to a V-shaped spectrum at 2.19 K. This is reminiscent of the temperature dependent gap measurements on the cuprates from chapter 4, where the normal state exhibits a similar V-shaped spectrum due to inelastic excitations. Above 2.19 K the spectra do not change significantly anymore which is why the superconducting transition is believed to have happened between 1.37 K and 2.19 K. This is generally in agreement with the transition temperature expected for this sample from the phase diagram in Fig. 5.1. One should keep in mind that this is a local measurement of the gap and variations from heat capacity or bulk transport measurements are expected.

5.5 *LC* Resonator Model for Plasmon Modes

That the weight for inelastic tunnelling events including n boson excitations is higher than for the emission of a single boson is surprising but the behaviour of $\alpha^2 F(\omega)$ reminds of the multiphoton-assisted tunnelling in superconducting junctions [192]. There, the ratio between microwave amplitude and photon energy yields a cut-off energy that suppresses multiphoton processes with an energy higher than this cut-off and at the same time it irregularly weighs the n multiphoton processes below the cut-off energy in a series of Bessel functions. The emergence of additional side-peaks in $\mathrm{d}I/\mathrm{d}U$ due to inelastic processes including the absorption and stimulated emission of photons in the microwave field is well described in the Tien-Gordon model [192]. Here, no microwave signal was coupled into the junction. Instead, the logic is reversed: In the absence of a microwave field, only spontaneous emission is possible, which is in agreement with the experimental observation that only high-energy side-peaks are present. The repeated emission of photons in the inelastic tunnelling process leads to energy level fluctuations on the time scale $1/\omega_{\mathrm{exc}}$ and with an amplitude U_{hf} that plays the role of the microwave intensity.

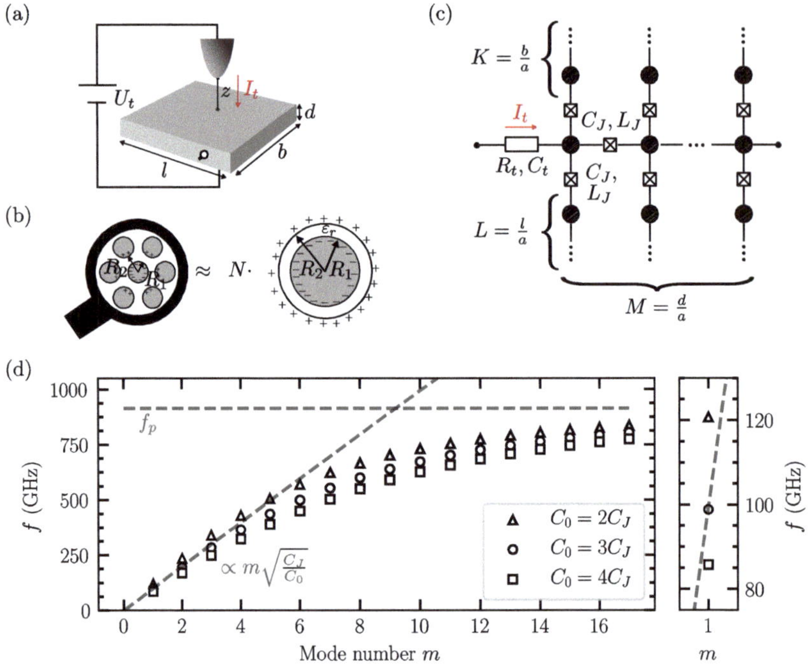

Figure 5.12: *LC* **Resonator Model**: (a) Schematic current flow through the grAl film in the STM. (b) Self-capacitance of each grain is approximated by a metallic sphere with radius R_1 in a dielectric shell of permittivity ε_r and thickness $(R_2 - R_1)$. (c) Electrodynamic model of the network of grains. The tunnelling current I_t flows through the central line of Josephson junctions in which each grain is capacitively coupled in parallel to other grains that lie in the same plane. (d) 2D plasmon modes for $m = 1, 2.., M$ and different in-plane capacitive couplings C_0. Right: Zoom-in on the mode $m = 1$.

Since the emission of the boson most likely happens inside the sample, one should not think of bare photons like in the vaccum. The sample is a dielectric, thus, the photons strongly couple to the charge degrees of freedom to form plasmons. Before we continue down this road and model the footprint of spontaneous multiplasmon emission processes in the tunnelling spectrum (Sec. 5.6), it is first time to consider a concrete model that legitimates the interpretation as plasmon modes. The main argument, here, will revolve around the energy of the mode.

Fig. 5.12(a) schematically shows how the grAl film is electrically wired in the STM geometry and in which direction the tunnelling current I_t flows. In our case, the measures of the film are $l = 2b = 10\,\text{mm}$ and $d = 50\,\text{nm}$. We adopt the electrodynamic model from Ref. [187] and describe the grAl film as a network of effective Josephson junctions (JJ). In the RCSJ (resistively and capacitively shunted junction) model [46, pp. 202], each JJ owns a self-capacitance C_J and a Josephson inductance L_J. In the Coulomb blockade regime, the aluminium grains act as individual capacitors. Let us assume that each grain is completely embedded in the dielectric with permittivity ε_r and closely surrounded by other grains which constitute the opposite electrode (Fig. 5.12(b)). We thus model the grain network as N metal spheres of radius R_1, each surrounded by a dielectric shell of width $(R_2 - R_1)$. The self-capacitance of such a spherical capacitor is then given by

$$C_J = 4\pi\varepsilon_0\varepsilon_r \frac{R_1 R_2}{R_1 + R_2}. \tag{5.3}$$

Evaluating this for $R_1 = 1.25\,\text{nm}, R_2 = 1.75\,\text{nm}$ and $\varepsilon_r = 9.0$ [193] yields $C_J = 0.73\,\text{aF}$. This value is very close to the capacitance determined through the charging peaks. The Josephson inductance can be determined through the critical current by the formula [30]

$$L_J = \frac{\Phi_0}{2\pi I_c}. \tag{5.4}$$

We use a critical current density of $j_c = 0.88\,\text{mA}/(\mu\text{m})^2$ for the film with DC resistance $\rho_{\text{dc}} \approx 2000\,\mu\Omega\,\text{cm}$ [187] and an effective current-carrying area of $A = a^2$, where $a = R_1 + R_2$ is the mean intergrain distance. With these values, we obtain $L_J = 42\,\text{nH}$. The effective plasmon frequency of one JJ,

$$f_p = \frac{\omega_p}{2\pi} = \frac{1}{2\pi}\sqrt{\frac{1}{L_J C_J}}, \tag{5.5}$$

is then $f_p = 914\,\text{GHz}$.

In Fig. 5.12(c), the wiring from Fig. 5.12(a) is translated into an effective circuit model. Each shaded blob depicts a single grain. Since the tunnelling current is injected in z-direction, the current only flows through a handful of grains, i.e. approximately $M = d/a$ effective JJs are put in series. But in the plane of the film, each grain is still capacitively coupled to its nearest neighbours. There are approximately $K = b/a$ grains in a volume $V = a^2 b$ and $L = l/a$ grains in a volume $V = a^2 l$. We again make use of the electrodynamic model from Ref. [187] and treat the circuit as quasi-1D. This approximation is valid due to the non-uniform electric field induced by the tip electrode. The electric field quickly falls off in the lateral direction of the film, i.e. in first order, the current only flows through the central line in Fig. 5.12(c). The capacitive coupling to neighbouring grains in each plane of the film (vertical connections in (c)), is captured in a mean self-capactiance of each plane C_0 and mainly determined by the average number of nearest neighbours. As demonstrated in Ref. [187], one obtains the energy levels of the quantized 2D plasmon modes by the equation

$$\omega_m = \frac{m\pi}{M}\left(L_J\left(C_0 + \frac{m^2\pi^2}{M^2}C_J\right)\right)^{-1/2} \tag{5.6}$$

with $m = 1, 2, .., M$ and $\omega_\infty = \omega_p = 1/\sqrt{L_J C_J}$ being the plasma frequency of the JJ. In Fig. 5.12(d), the energies of the discrete 2D plasmon modes due to confinement in z-direction are shown for $C_0 = 2C_J, 3C_J, 4C_J$, i.e. two, three or four nearest-neighbour grains in each plane. With $f_p = 914\,\mathrm{GHz}$, the bulk plasmons are roughly ten times higher in energy than the bosonic mode we see in our tunnelling spectrum. The first standing wave mode ($m = 1$) due to confinement by the film thickness d, however, lies between $f_1 = 86$–$121\,\mathrm{GHz}$ for $C_0 = 2$–$4C_J$. This corresponds to $\omega_1 = 355$–$499\,\mu\mathrm{eV} \approx 1.28$–$1.80\,\Delta$, which is in the range of our observed ω_{exc}. It should be noted that this is only a rough estimation. The large uncertainty of the critical current density [187] is the major factor in error estimation of ω_p, as it only enters L_J. The grain size plays a secondary role, as L_J scales inversely but C_J linearly with the cross-section, such that the dependence nearly cancels out. Additionally, the good agreement between the C_J estimated from the microscopic model and the value deduced from the charging peaks give reason to believe that the estimate for C_J is more accurate than for L_J.

In the following section, the spontaneous multiplasmon emission process is simulated in the Tien-Gordon model. Circumstances under which the excitation of plasmons are easy would be consistent with a low intergrain coupling and low superfluid stiffness, i.e. the system is highly susceptible to phase fluctuations of the order parameter near the SIT [149, 184].

5.6 Tien-Gordon Model for Spontaneous Plasmon Emission

The Tien-Gordon model [192] traditionally describes photon-assisted tunnelling in superconducting junctions that are exposed to a radiation field $U_{\mathrm{hf}} \cos(\Omega t)$. The electrons can tunnel inelastically via the spontaneous absorption and stimulated emission of photons with the energy Ω. The resultant tunnelling spectra exhibit n sidebands corresponding to the absorption/emission of n photons of energy Ω [194]. How many sidebands occur, depends on the amplitude of the applied field U_{hf}. Usually, spontaneous emission is neglected because the rate for stimulated emission is much higher in the radiation field.

In this work, no high frequency signal was coupled into the tunnel junction. Instead, the system is considered to be inherently susceptible to inelastic tunnelling events that include the spontaneous emission of n photons in the superconductor. Inside the superconductor, the coupling between charge carriers and photons leads to collective excitations, the plasmons. Via the Anderson-Higgs mechanism, these

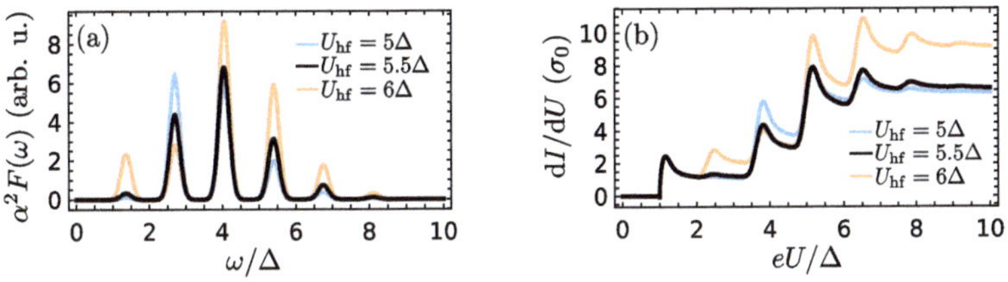

Figure 5.13: Spontaneous Plasmon Emission: The effective tunnel Eliashberg Function $\alpha^2F(\omega)$ (a) and the differential conductance $\mathrm{d}I/\mathrm{d}U$ (b) are shown in the analytic description of the Tien-Gordon model considering only plasmon emission. For $5\Delta \lesssim e|U_{\rm hf} \lesssim 6\Delta$, inelastic processes with up to $n = 5$ plasmons contribute significantly to the tunnel current and $n = 3$ is the process with the largest weight (cf. peak height in (a)).

plasmonic excitations also couple to the superconducting phase. It is assumed that the spontaneous emission can, in this case, be described analogously to the spontaneous absorption by simply adjusting the energy conserving term in the δ-distribution. A more rigorous treatment of this problem by the authors of Ref. [195] shows that the total probability for a spontaneous n plasmon emission is described by the convolution of two distribution functions: $P_0(E)$ and $P_{\rm occ}(E)$. P_0 describes the probability that the tunnel junction emits a photon into the vacuum and $P_{\rm occ}$ describes the absorption capability of the "cavity", which in this case is the grAl film. If the cavity is driven into a coherent state $|\alpha\rangle$ with amplitude α_0, then $P_{\rm occ}$ yields the Tien-Gordon distribution. Additionally, one realizes that the classical AC voltage amplitude $U_{\rm hf}$ is proportional to the amplitude of the plasmon excitation $|\alpha_0|$ [195]. We thus consider our inelastic tunnelling current to behave like a classical drive to the resonating aluminium grains.

Neglecting absorption, i.e. considering only the high-energy sidebands, and restricting ourselves to a single plasmon mode with energy $\omega_{\rm exc}$, the total conductance for single quasiparticle tunnelling in the Tien-Gordon model is written as

$$\begin{aligned}\sigma_{\rm s}^{\rm tot}(eU) &= \int_0^{eU} \mathrm{d}\omega \sigma_{\rm s}^{\rm el}(eU-\omega) \sum_n J_n^2\left(\frac{eU_{\rm hf}}{\omega_{\rm exc}}\right)\delta_a(\omega - n\omega_{\rm exc}) \\ &= \sigma_{\rm s}^{\rm el}(eU) + \underbrace{\int_0^{eU} \mathrm{d}\omega \sigma_{\rm s}^{\rm el}(eU-\omega) \sum_{n=1} J_n^2\left(\frac{eU_{\rm hf}}{\omega_{\rm exc}}\right)\delta_a(\omega - n\omega_{\rm exc})}_{\propto \alpha^2 F(\omega)}\end{aligned} \tag{5.7}$$

with

$$\delta_a = \frac{1}{|a|\sqrt{\pi}} \mathrm{e}^{-(x/a)^2}. \tag{5.8}$$

J_n are Bessel functions of order n and $U_{\rm hf}$ is proportional to the amplitude of the plasmon state. The Bessel function for $n = 0$ describes elastic tunnelling. Hence, the terms with $n > 0$ can be collected to an effective "tunnel Eliashberg function" $\alpha^2 F(\omega)$.

In Fig. 5.13, the function $\alpha^2 F(\omega)$ and the resulting differential conductance in the Tien-Gordon model is shown for $\omega_{\text{exc}} = 1.35\Delta$, $\sigma_{\text{s}}^{\text{el}}(eU) = \sigma_0|eU|/\sqrt{(eU)^2 - \Delta^2}$, $a = 0.2\Delta$ and the three indicated values for U_{hf}. All $\alpha^2 F(\omega)$ were normalized by the condition of the elastic conductance

$$\sigma_{\text{s}}^{\text{el}}(eU) = J_0^2\left(\frac{eU_{\text{hf}}}{\omega_{\text{exc}}}\right)\int_0^{eU} \mathrm{d}\omega\sigma_{\text{s}}^{\text{el}}(eU-\omega)\delta_a(\omega) \overset{eU\to\infty}{\to} \sigma_0. \tag{5.9}$$

The ratio $eU_{\text{hf}}/\omega_{\text{exc}}$ yields a cut-off for the number of plasmons that are excited in one inelastic tunnelling event. For $eU_{\text{hf}} \lesssim 6\Delta$ the maximum number is $n = 5$. Due to the functional form of the Bessel functions, the ratio $eU_{\text{hf}}/\omega_{\text{exc}}$ also adjusts the relative amplitudes for the multi-plasmon excitations.

As can be seen from Fig. 5.13, good agreement with the experimental data, i.e. the process $n = 3$ has the largest weight, is found for $5\Delta \lesssim eU_{\text{hf}} \lesssim 6\Delta$. The resulting $\mathrm{d}I/\mathrm{d}U$ spectrum in the Tien-Gordon model then closely resembles the experimental spectrum with the characteristic excitation ladder up to a cut-off at $e|U| \approx \Delta + 5\cdot\omega_{\text{exc}}$.

5.7 First Results on Oxidized Polycrystalline Al

The STS measurements of the YSR states in oxygen-rich grAl (sample G1) from section 5.3 suggested that it is likely that the unpaired spins reside in the nonstoichiometric oxide of aluminium. In order to test this hypothesis, the pure and surface clean aluminium film (sample A) from section 5.2.1 was oxidized by exposure to a pure O_2 atmosphere. Oxygen gas (purity$= 99.9999\,\%$) was introduced to the UHV chamber through a variable leak valve up to a partial pressure of $p_{O_2} = 1\times10^{-5}$ mbar and th sample was oxidized for 1 h at room temperature. This corresponds to a dose of roughly 2.7×10^4 L (Langmuir). Due to the quick passivation of the aluminium surface after the initial incorporation mechanism (the first chemisorbed monolayer of O_2 is buried in the second surface layer by updiffusion of aluminium), the oxygen uptake rises only slowly for higher doses [196] and saturates after roughly 2 hours [197]. The exposure dose used here should lead to no more than 2 ML of AlO [197–199]. If dangling bonds in grain boundaries and interfaces of the oxide are the host for unpaired electrons, then YSR states should also be observed at the surface of the oxidized aluminium film. Conveniently, complicated factors like size effects of the Al grains, disorder of the superconductor and percolation effects are taken out of the equation in this control experiment.

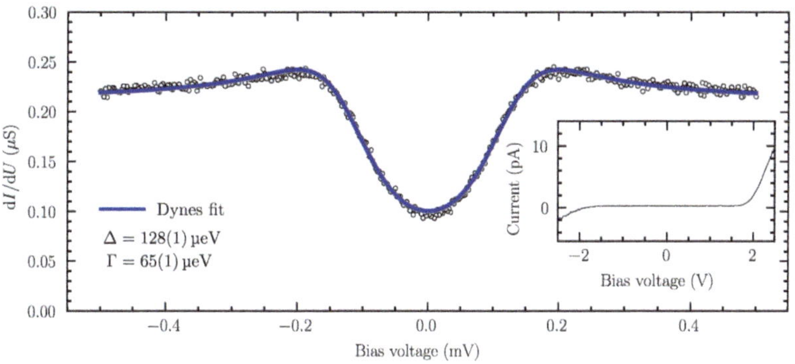

Figure 5.14: Oxidized Al Film: dI/dU spectrum with superconducting gap obtained on the strongly oxidized Al film at $T = 39\,\mathrm{mK}$. No YSR states are visible. The blue line shows a Dynes DOS fit with $\Delta = (128 \pm 1)\,\mu\mathrm{eV}$ and quasiparticle scattering rate $\Gamma = (65 \pm 1)\,\mu\mathrm{eV}$. Feedback condition: $U = 0.5\,\mathrm{mV}$, $I = 100\,\mathrm{pA}$. Inset: Large bias range $I(U)$ curve displaying the low conductance of the junction. Due to the large insulating gap the STM tip is in contact with the surface for low bias voltages. The tunnel barrier is the aluminium oxide not the vacuum.

Fig. 5.14 shows the dI/dU spectrum of the oxidized aluminium film at $T = 39\,\mathrm{mK}$. The superconducting gap is reduced and not fully developed. The coherence peaks are also broadened significantly. In-gap states are not visible. The modified appearance of the superconducting gap is caused by the fact that tunnelling into the superconducting aluminium is only made possible by bringing the STM tip in contact with the surface oxide. In contrast to the vacuum, the oxide layer allows not only ballistic but also diffusive transport. That means that the tunnelling electron can in principle inelastically scatter on defects inside the oxide. This form of dissipation is seen as excess sub-gap differential conductance and absent coherence peaks. The exponential decay of the tunnelling current makes it virtually impossible to record a tunnelling current if the tip is free standing above the surface because an additional insulating layer of $\approx 0.8\,\mathrm{nm}$ is separating it from the superconducting aluminium. The good insulation of the thin surface oxide is well captured in the large range $I(U)$ curve in the inset of Fig. 5.14. Only at a stabilization voltage of $U > 2\,\mathrm{V}$ a tunnelling current of few pA starts to flow. To account for scattering processes of electrons with sub-gap energies in the presence of the oxide, the superconducting gap was fitted to a Dynes DOS function (see Chap. 4 for a functional expression). The fit yields $\Delta = (128 \pm 1)\,\mu\mathrm{eV}$ and a scattering rate of $\Gamma = (65 \pm 1)\,\mu\mathrm{eV}$. If YSR states are not present or just overshadowed by the poisoning scattering channels for sub-gap energies cannot be discerned.

This seemingly simple control experiment beared unexpected complications due to the quick evolution of a large insulating gap of roughly $4\,\mathrm{eV}$, already for a supposedly thin oxide layer. A great amount of preparation of a tip with good energy resolution (in order to resolve the YSR states) is opposed to its one time use. As soon as the tip is dipped into the oxide, the measurement has to be performed at once. At each lift off the surface the tip picks up a cluster of the insulating oxide, which successively increases the effective tunnelling barrier for the next spectroscopy until no tunnelling current can be detected anymore. It is expected that the measurement of an oxidized Al film with a thickness of $0.5\,\mathrm{ML} < \Theta < 1\,\mathrm{ML}$ yields more conclusive results concerning the existence of YSR states at the AlO surface. To study the depencence between oxide thickness and abundance of these states, however, the experimental procedure would have to be adjusted to eliminate the problems encountered in the tunnelling geometry chosen here.

Density functional theory (DFT) calculations of chemisorbed O_2 on a few layers of $\alpha-Al_2O_3$ on Al(111) predict that the charge transfer of the oxygen molecule heavily depends on the oxide thickness and an effective spin of $1\,\mu_B$ on the adsorbant is expected only at $4-9$ trilayers of Al_2O_3 [200]. In order to realize such thick aluminium oxide layers either much larger oxygen pressures [199] or elevated temperatures that facilitate the surface reactivity (dissociative chemisorption of O_2 is thermally activated) [201] are needed. The latter method would be the preferred in a UHV environment like the setup used in this work.

5.8 Summary

The combined STM/STS measurements on grAl in the low resistivity and high resistivity regime confirmed the existence of nano-sized grains and yielded enhanced, uniform superconducting gaps of $\Delta = (298 \pm 1)\,\mu\mathrm{eV}$ and $\Delta = (312 \pm 1)\,\mu\mathrm{eV}$ compared to $\Delta = (191 \pm 1)\,\mu\mathrm{eV}$ for a polycrystalline aluminium sample. The oxygen-poor grAl sample displayed metallic behaviour for larger bias voltages, which speaks for well coupled grains, and the superconducting gap was free of sub-gap states. In the oxygen-rich grAl sample, charging effects, indicative of the Coulomb blockade regime, were observed. Apparently, the grains decouple as a function of oxide thickness and can be collectively charged in the form of clusters containing several electrically well connected grains. In some grains, YSR states were found as a fingerprint of localized spins. The energetic shift of the YSR states in dependence of an applied electric field suggests that the unpaired spins reside in the oxide barriers. An independent STS experiment on a polycrystalline aluminium film with an estimated surface oxide thickness of $\approx 2\,\mathrm{ML}$ revealed no YSR states. It is likely that the localization of unpaired spins requires a thicker oxide. On individual grains in the oxygen-rich grAl sample, the out-gap structure of scanning tunnelling spectra is shaped by inelastic excitations with long lifetimes. The extracted effective bosonic spectral density of these excitations can be qualitatively well explained by the spontaneous emission of multiple plasmons. This would very much support an SIT that is driven by phase fluctuations of the superconducting gap. From an energy standpoint, we argued that the excited plasmon mode is most likely part of the lowest branch of discrete modes that are quantized through the thickness of the grAl film. An alternative, yet very speculative, candidate for these inelastic excitations are Higgs modes.

6 Multiflux Vortex States in Bulk Pb

The recent evidence for two-band superconductivity [39] and its naturally close proximity to the Bogomolnyi point [41, p. 436] make lead a prime candidate to study the interplay of multiband superconductivity and normal conducting domains in the intermediate state. Bulk multiband superconductivity in conjunction with conventional s-wave pairing has been proposed to unveil a new route to topological superconductors that are robust to magnetic fields and host Majorana modes [202, 203]. At medium magnetic fields, the low-temperature transitional region of a multiband superconductor with close to critical κ, is also predicted to host interesting structures like multiflux vortices [204]. These vortices represent topological objects themselves, that may also host Majorana modes under the right conditions [205]. In this chapter, new findings on the appearance of the low-temperature intermediate state of Pb are presented. These include the emergence of single-flux vortices and multiflux vortices. In the process, the influence of interband coupling, tunnelling into Caroli-de Gennes-Matricon states and a robust method for winding number determination are detailed. The numerical simulation of the vortex patterns via solution of 3D Eilenberger equations confirms the interpretation of presented scanning tunnelling data. The Fermi velocities that enter the numerical simulations were provided by Rolf Heid from IQMT, KIT, and are obtained within a fully relativistic DFT approach [206]. This chapter is based on the following preprint: [207].

6.1 Meißner and Intermediate State

After several cycles of hot sputtering and annealing, as described in Sec. 3.3.2, only very few adsorbants remained at some kink sites of step edges on the Pb(111) surface and terraces formed with widths of up to few hundred nanometres, as can be seen in the topographic scan shown in Fig. 6.1(a). Upon zero-field cooling the Pb(111) sample to 39 mK, it enters its superconducting state below $T_c \sim 7.2\,\mathrm{K}$ [41, p. 436]. The $\mathrm{d}I/\mathrm{d}U$ spectrum in Fig. 6.1(b) demonstrates that the two superconducting gaps of Pb can be resolved in our setup, even with a normal conducting tip. From the fit of a temperature broadened two-gap BCS function (red curve), the gaps are

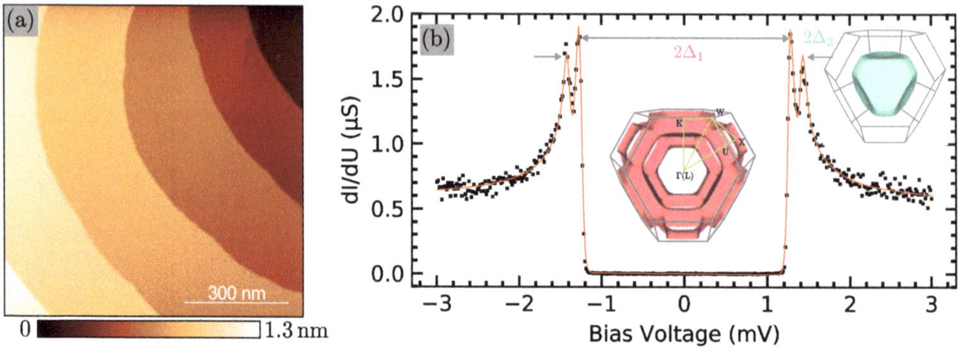

Figure 6.1: Superconducting State: (a) Topographic scan of the Pb(111) surface showing almost no contamination and wide terraces. (b) Differential conductance spectrum at $T = 39\,\mathrm{mK}$ centred around the Fermi energy. The superconducting gaps $\Delta_{1,2}$ are determined from a broadened two-gap BCS function fit (orange curve). The insets show 3D models of the FSs of the two superconducting bands (from Ref. [209]) inside the first BZ (wireframe) viewed from the (111) direction. Δ_1 is assigned to the band forming a tubular FS (red). The larger gap Δ_2 is assigned to the band with a compact FS (cyan).

determined as $\Delta_1 = (1.26 \pm 0.02)\,\mathrm{meV}$ and $\Delta_2 = (1.40 \pm 0.02)\,\mathrm{meV}$. This is in good agreement with previous measurements of the difference of the two gaps [39], where the authors made use of the energy resolution amplification of superconducting tips.

It should be noted that the intensity difference of the two coherence peaks has previously been attributed to the k-dependent tunnelling matrix element and the smaller (larger) gap has been assigned to the compact (tubular) Fermi surface (FS) sheet [39], shown in the cyan (red) 3D model in Fig. 6.1(b). This is in contrast to BdG calculations [208] and the findings in this work. A conclusion from the analysis in Sec. 6.2.2 is that the band-to-gap assignment should be reverse: The band with the tubular/compact FS (red/cyan) is from here on denoted as band I/II and is responsible for the superconducting gap Δ_1/Δ_2.

After applying a perpendicular magnetic field of $B = 85\,\mathrm{mT}$, which is above the critical field of $\mu_0 H_c \sim 80\,\mathrm{mT}$ [210], magnetic flux enters the sample from the sides and destroys superconductivity completely. Upon subsequent decrease of the field below H_c, the Landau intermediate state is reached. This state is characterized by large normal and superconducting domains and by long domain walls separating the two. The shapes and sizes of domains in the intermediate state of lead have been extensively studied by magneto-optical methods revealing the strong dependencies on temperature, sample shape and magnetic protocol [211–214]. Since the domain sizes are in the range of several micro- or even millimetres, STM is not exactly suited to study these domains. Still, it is possible to frequently freeze domain walls in the scan frame by performing a slow magnetic field ramp ($5\,\mathrm{mT/min}$) and simultaneously measuring the differential conductance at $eU_t \sim \Delta_2$ at the tip position. As soon as a

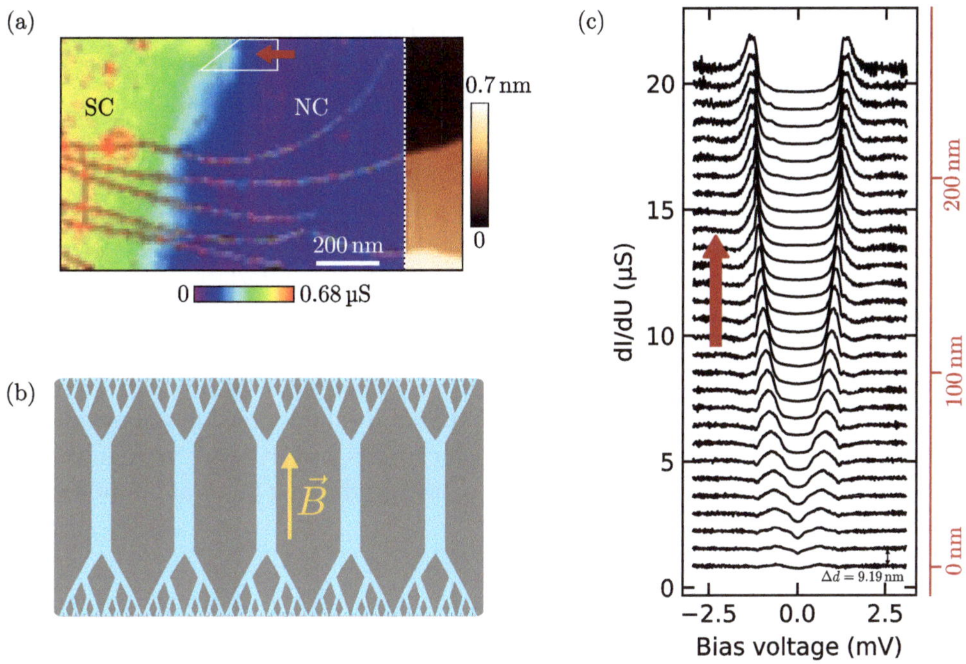

Figure 6.2: Intermediate State: (a) $\mathrm{d}I/\mathrm{d}U$ map at $U_t = 1.3\,\mathrm{meV}$ showing a typical domain wall in the intermediate state at $B = 23\,\mathrm{mT}$. The normal (superconducting) domain exhibits a low (high) conductance at this bias voltage. Right overlay: Corresponding topographic scan image of the area. (b) Schematic cross-sectional view through the bulk material in the intermediate Landau state. The magnetic flux density (blue) is favouring a more homogeneous distribution at the surface than in the bulk leading to branching. (c) $\mathrm{d}I/\mathrm{d}U$ spectra along a cross-section from normal conducting to superconducting domain indicated by the white frame in (a). The direction of the profile is indicated by the red arrow. The spectra are locally averaged over a straight part of the domain boundary (perpendicular to the profile direction) and recorded in distance increments of $\Delta d = 9.19\,\mathrm{nm}$.

domain wall passes the tip position, the measured conductance jumps and the ramp is stopped. After a few seconds of waiting until the dynamics have stopped, finding a domain wall within a scan frame of $1.4\,\mu\mathrm{m} \times 1.4\,\mu\mathrm{m}$ is likely.

A typical domain wall in the intermediate state is shown in the $\mathrm{d}I/\mathrm{d}U$ map in Fig. 6.2(a). At a tunnelling bias of $U_t = 1.3\,\mathrm{mV}$, the normal conducting domains show up as areas of low conductance (purple) and the superconducting domains as areas of high conductance (green/yellow). Thin lines of high conductance are an artefact of the measurement mode and appear at the position of step edges on the surface as the right overlay shows. They are the consequence of a sudden jump in the tunnelling current when the tip gets too close to an edge in the multipass mode. They are therefore ignored in the discussion of the LDOS. A cross-sectional line scan across the domain wall, as in Fig. 6.2(c), shows how the gap Δ_1 changes from zero to its maximum on the length scale of the coherence length. The local recovery of superconductivity agrees well with reported coherence lengths of $\xi \sim 87\,\mathrm{nm}$ [41, p. 436]. A detailed analysis of the coherence length $\xi_{1,2}$ of the two bands, measured inside vortices, will follow in Sec. 6.2.1. Direct information about the magnetic field lines and therefore also the local supercurrents inside the superconductor bulk cannot be gained from neither STM nor the magneto-optical (MO) method mentioned earlier. From energetic considerations, however, the intermediate state is supposed to form according to two competing effects: The interface energy to build domain walls is minimized by a course distribution of magnetic flux in the interior, while the magnetic field outside the sample favours homogeneity. The resulting compromise is a branching flux pattern, as shown in Fig. 6.2(b), which was first predicted by Lev Landau [215, 216]. These thermodynamic considerations neglect the influence of flux pinning centres in real crystals and dynamic effects caused by the magnetic field ramp, Lorentz force acting on the flux tubes and topological hysteresis [212].

6.2 Emergence of Isolated Vortices

For disk-shaped Pb samples, R. Prozorov could show that the equilibrium state after flux penetration of a zero-field cooled (ZFC) sample is tubular: With increasing field, flux enters the sample from the outer edges and is quickly pushed to the centre by the Lorentz force. The tubular structure grows with increasing field in a coarsening froth pattern. At the same field values, the sample exhibits a laminar magnetic flux pattern on the surface (in accordance with Landau's prediction) if this state has been reached from the normal state upon field decrease [212]. While this is an experimental observation on the micrometer scale and at 5 K, it is nevertheless highly unexpected that isolated vortices, so flux tubes carrying a single flux quantum, form in Pb at $\sim 30\,\mathrm{mK}$ when the field is decreased following the magnetic protocol described in the previous section. If single-flux vortices are indeed stable in a bulk Pb single crystal, then it raises the question whether strong pinning centres can explain this fact, or the classification of superconductors into type-I and type-II has to be revisited at ultra-low temperatures $T \ll T_c$.

6.2.1 Normal Single-Flux Vortices

Here, compelling evidence for the existence of stable magnetic vortices in intermediate state bulk Pb(111) is shown by display of the local density of sub-gap states in a magnetic vortex, called Caroli-de Gennes-Matricon (CdGM) states. The quantum mechanical nature of theses states is explained in detail in Sec. 6.3. For now, the following quasi-classical picture suffices to discuss most features of the experimentally observable LDOS: In a magnetic vortex core, the superconducting order parameter or gap size vanishes. Normal state electrons and holes with $E < \Delta$ are trapped in this vortex as they are (mostly) specularly reflected by the potential barrier Δ. Thus, a great number of quantum well states (QWS) exist inside the magnetic vortex core. Their abundance at a certain energy is determined by the normal state DOS and their penetration depth into the superconducting surrounding is determined by the Fermi velocity $\boldsymbol{v}_F$[1].

After ramping the magnetic field down from $85\,\mathrm{mT}$, isolated, round normal conducting domains of only $\sim 100\,\mathrm{nm}$ in diameter are frequently stabilized. They appear as a conductance depression at $eU_t = \Delta_2$, as displayed in Fig. 6.3(h). The appertaining zero bias conductance map is shown in Fig. 6.3(a). It reveals a threefold symmetric star-shaped pattern with a maximum in the star's centre. The CdGM LDOS stretches over $100\,\mathrm{nm}$ in the $\langle 2\bar{1}\bar{1}\rangle$ directions in the form of thin rays (blue/green) that continuously fade at increasing distances from the centre. On top of that, three weak star arms (dark blue) along the $\langle 1\bar{2}1\rangle$ directions are visible. The bulk crystal directions have been determined from larger crystal defects that reveal the direction of glide planes. Further information can be found in Chap. 7.

With increasing energy (independent of sign of bias voltage), the star's weak and strong arms both split into two with increasing splitting distance (see Fig. 6.3(b-f)). Meanwhile, the central peak splits isotropically to a ring shape (Fig. 6.3(b-f)). At $eV \lesssim \Delta_1$ the ring size is maximal (see Fig. 6.3(g)). This pattern proved to be invariant under reversal of tunnelling current or magnetic field direction, which is demonstrated in Appendix D and showcases the particle-hole symmetry of the CdGM states.

By solving the quasi-classical 3D Eilenberger equations, the experimentally found LDOS patterns of CdGM states could be qualitatively reproduced for bands I and II separately. In Fig. 6.3(a,e), the right bottom (top) panel shows the calculated LDOS signature for a single-flux vortex using the DFT calculated band structure from band I (II) at given energy. Apparently, the long, stretched-out star arms of high LDOS stem from electrons at the Fermi edge of band II, for which the FS is shown in cyan. A view onto the FS from the (111) direction makes it obvious, that these long arms are caused by the large Fermi velocity anisotropy in this band. A large portion of the FS consists of flat parts where the Fermi velocity's in-plane component points in one

[1] For a non-relativistic electron, the classical momentum $\boldsymbol{p} = m\boldsymbol{v}$ can be used in the kinetic energy term of the Hamiltonian $H_{\mathrm{kin}}(\boldsymbol{r}) = -i\hbar\boldsymbol{v}\nabla$.

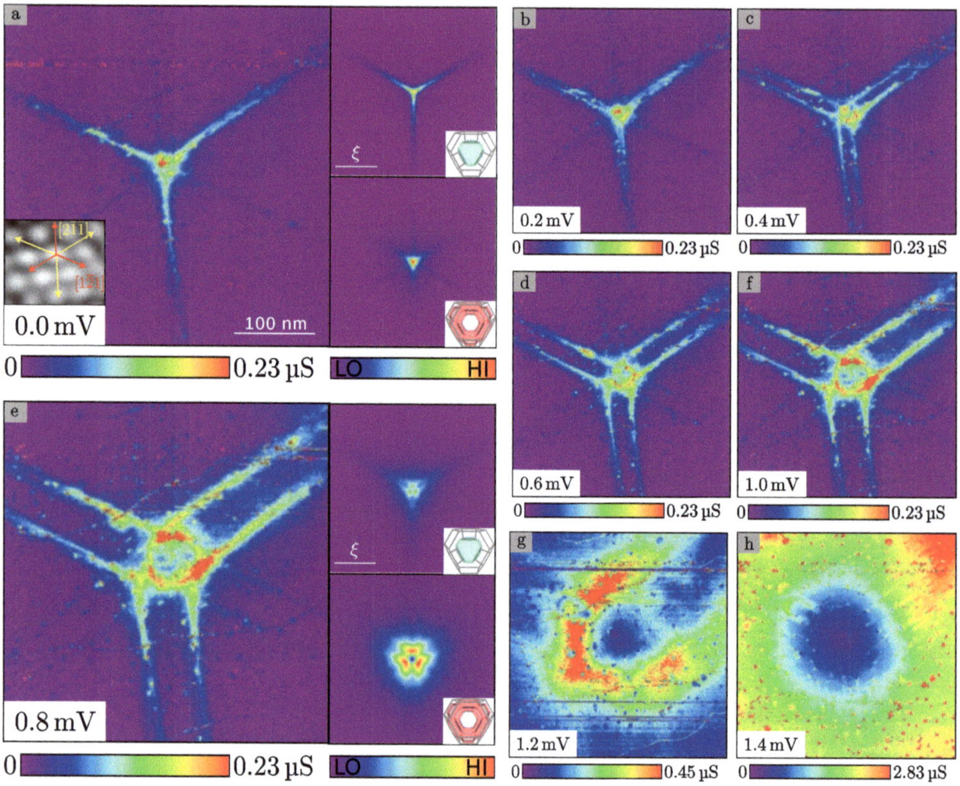

Figure 6.3: Single Flux Vortex: Multi-pass images of the differential conductance at sub-gap energies (a-h), recorded at $T = 37\,\mathrm{mK}$ and $B_{\mathrm{ext}} = 0\,\mathrm{mT}$. The inset in (a) shows the atomic lattice and important bulk crystal directions are marked by yellow/red arrows. Right panels in (a,e) complement experimental data with LDOS simulations using a 3D Eilenberger formalism and DFT calculated Fermi velocities. The LDOS pattern was calculated seperately for band I (top) and band II (bottom) with FSs as shown in the small inset.

of the $\langle 2\bar{1}\bar{1}\rangle$ directions. The splitting of the arms is also seen in the semi-classical calculation at non-zero energy. The root of this effect lies in the quantum-mechanical nature of these states and will be discussed in detail in Sec. 6.3. The calculations for band I confirm that the weak arms in the $\langle 1\bar{2}1\rangle$ directions as well as the strong central LDOS peak stem from electrons at the red FS. The openness of this FS provides for a larger variety of Fermi velocity direction, which effectively leads to a more isotropic scenario, albeit the threefold symmetry still remains apparent. The stronger confinement of states in the centre is a result of the (on average over the whole FS) larger z-component[2] of the Fermi velocity. Consequently, the respective

[2] Here, the z-direction denotes the direction perpendicular to the crystal surface, so the [111] direction.

in-plane component is smaller, which leads to a smaller penetration depth into the superconducting surrounding. The out-of plane component of the Fermi velocity proved to play a crucial role in the Eilenberger calculations, as solving effective 2D Eilenberger equations in the (111) plane would only yield sixfold symmetric LDOS patterns. The 3D trajectories of the quasiparticles do, however, have substantial z-components. Upon respecting the translation symmetry breaking by the surface, in conjunction with 3D quasiparticle trajectories, the threefold symmetric LDOS patterns could finally be reproduced. A detailed explanation of the simulation methods used, can be found in Appendix C.

It should be noted that this is not the first observation of anisotropic LDOS signatures due to CdGM states in and around a vortex structure: The first to observe such star-shaped structures were Hess et al., who observed them in the vortex lattice of 2H-$NbSe_2$ [217]. Later, Hayashi et al. demonstrated that the semi-classical theory of Eilenberger can reproduce the experimental results nicely [218, 219]. Recently, they were also observed on the (0001) surface of the elemental superconductor La [220], where the anisotropy was attributed to the Fermi velocity and not to an anisotropy in the gap parameter which was shown to have a large affect the LDOS pattern, too [221, 222]. This work marks the first experimental observation of highly anisotropic CdGM states from two different bands inside solitary vortices. The early stated presumption that the anisotropic pattern might be caused by neighbouring vortices [217, 223] in a flux lattice can therefore be ruled out. In addition, this is the first time, a pattern with odd-integer rotational symmetry is found. It could be shown that this pattern can be reproduced using 3D Eilenberger simulations with accurate Fermi velocites and that the total signature can be understood as a superposition of the CdGM states from both superconducting bands.

After learning which sub-gap states belong to which band, it makes sense to take a deeper look into the single bias spectroscopies inside the vortex in order to determine the length scale on which Δ_1 and Δ_2 change. Previously, in Sec. 6.1, this proved to be difficult for the domain wall due to the ripples inside it[3]. Radial linegrids recording the differential conductance were taken on the vortex from Fig. 6.3. In Fig. 6.4, the second derivative of the differential conductance $-\partial^2\sigma/\partial V^2$ is shown for four different directions (indicated in the inset). The grey color code has been adjusted such that white regions mark clear peaks in the differential conductance curves. One can now follow the traces of these peaks as the distance from the centre increases. For $\alpha = 0°$ the linegrid is aligned with a strong star arm. We can thus recognise the trace of peaks at zero energy to be the CdGM states of band II (cyan

[3] The necessary averaging over a certain length of the domain wall had a negative impact on the overall energy resolution and made a distinction between Δ_1 and Δ_2 difficult over a large distance.

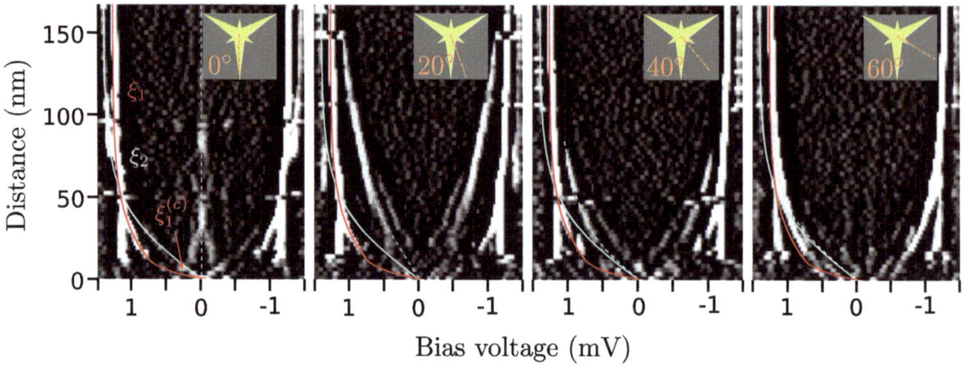

Figure 6.4: Coherence Lengths in the Normal Single-flux Vortex: Angle dependent radial measurement of the vortex core states from the vortex in Fig. 6.3. Displayed is the second derivative of the differential conductance $-\partial^2\sigma/\partial V^2$ in order to highlight maxima in the LDOS. The point distance of single spectra is $\Delta d = 2.57\,\mathrm{nm}$. Marked are the opening of Δ_1 (red line), the opening of Δ_2 (cyan line) and the CdGM states of band II (cyan dashed line). The inset shows the direction of the y-axis.

dashed line). For $0° < \alpha < 60°$ the states move to higher energies and eventually fall together with the opening of Δ_2 at $\alpha = 60°$ (cyan line). A crossing of $\Delta_1(r)$ and $\Delta_2(r)$ is identified at $d \approx 50\,\mathrm{nm}$. That means, that the length scales on which the two recover to their maximum value must be different. The smaller gap Δ_1 seems to recover much quicker than Δ_2 when moving away from the vortex centre (red line). While $\Delta_2(r)$ can be described by $\Delta_2(r) \sim \Delta_2 \tanh(r/\xi_2)$ with $\xi_2 \sim 45\,\mathrm{nm}$ over the whole distance, Δ_1 cannot. It follows a tanh behaviour with $\xi_1 \sim 45\,\mathrm{nm}$ for large distances from the centre but has a much steeper slope yielding a core size of only $\xi_1^{(c)} = \Delta_1(\infty)\left[\lim_{r\to 0}\frac{\mathrm{d}\Delta(r)}{\mathrm{d}r}\right]^{-1} \approx 10\,\mathrm{nm}$.

Such a shrinking of the vortex core size is known from theory as the Kramer-Pesch effect [223, 224]. It is so far mainly studied in isotropic vortices [224–226] and the self-consistent solution of the pair potential with isotropic Fermi velocity shown in Fig. D.2 confirms that a Kramer-Pesch effect should be present and lead to a significant deviation from a tanh shape of $\Delta(r)$. This observation is in accordance with our band-gap assignment: Band I, which is responsible for Δ_1, has a more isotropic Fermi velocity distribution than band II, for which the Kramer-Pesch effect is missing. It makes sense that, in the anisotropic case, the thermal population of the CdGM states at $T \ll T_c$ does not lead to a core size reduction because the low energy CdGM states can extend far from the vortex centre in certain directions. For a superconductor in the clean limit, the core size should shrink proportional to T/T_c and should even be able to become infinitely small at $T \to 0\,\mathrm{K}$ [227]. This raises the question, how the opening of the gap should rightfully be determined in the low temperature experiment. As one can see in Fig. 6.4, the quasiparticle coherence peak at Δ_1 is already present at $\approx 10\,\mathrm{nm}$ distance from the centre which could mean that the core size is even smaller and the red line follows the CdGM states of Δ_1 which do not coincide with the "opening" of Δ_1. This is, however, almost a philosophical question: Since these CdGM states are always present in a vortex, they always take away from the condensate and only when they reach $E = \Delta$ is the corresponding coherence peak maximally large and the gap fully empty. Apparently, there is no gap size matching (the similar Δ are rather a coincidence) and no length scale locking present in Pb. This leads to the conclusion that the bands are relatively decoupled from each other [87]. This readily justifies the separate treatment of the two bands in the quasi-classical simulation.

6.2.2 Anomalous Vortices

Apart from normal single-flux vortices, anomalous vortices like the one in Fig. 6.5 are frequently observed. These vortices also carry one single flux quantum and behave like the normal vortices but the sets of CdGM states from the two bands seem to be spatially displaced with respect to each other. The fact that the strong central zero bias peak in Fig. 6.3(a) belongs to a different band than the strong star arms was previously only inferred from the comparison with the theoretical calculations but can now be confirmed experimentally. In the anomalous vortex, this strong zero bias peak does not lie in the centre of the star pattern, as can be seen in Fig. 6.5(a). In Fig. 6.5(b-g), it is also much clearer that this peak splits almost isotropically to a ring-like pattern for higher energies. Due to the displacement of the CdGM states, the zero energy LDOS in the star centre and the ring centre can be compared. This is shown in Fig. 6.5(i). The amplitude of the tunnelling conductance in the ring centre is roughly four times larger than in the star centre. In fact, the band-gap assignment is unambiguously possible by using the fact that the CdGM states from different bands are spatially displaced: By recording the differential conductance in fine linegrids (Fig. 6.5(j,k)) along the paths k,j, that are marked in Fig. 6.5(i)'s right panel, one sees that Δ_2 (cyan line) closes in the star centre, while Δ_1 (red line) closes in the centre of the ring. Since the CdGM state-band assignment is known from the quasi-classical simulations, one can now also definitively relate band I (tubular FS) to the smaller Δ_1 and band II (compact FS) to the larger Δ_2.

The direction and magnitude of spatial displacement between the two state sets could not be tied to any crystal direction and appears to be almost random. By changing the magnetic field, the displacement can be modified, as Fig. 6.6 demonstrates. In fact, both CdGM state sets are mobile and as a result, the anomalous vortex can be transformed back into a normal vortex. As a point of proof, Fig. 6.7(a,c) shows the respective topographic images to the conductance maps in (b,d). Clearly, the scan

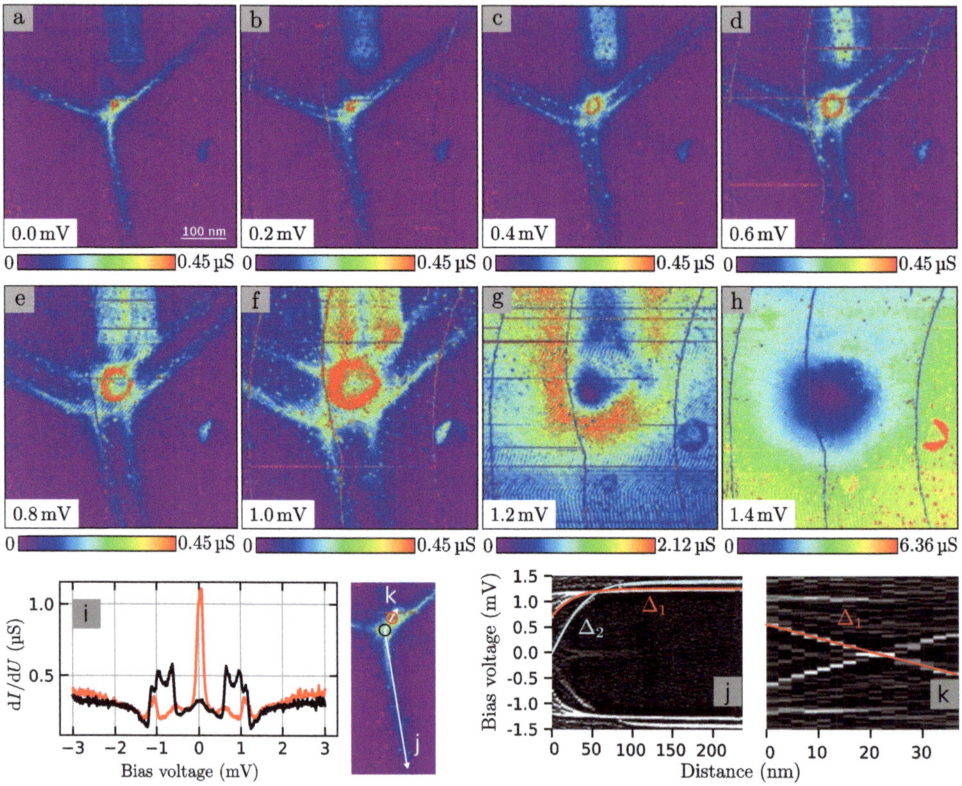

Figure 6.5: Anomalous Single-flux Vortex: (a-h) $\mathrm{d}I/\mathrm{d}U$ maps like in Fig. 6.3 for an anomalous vortex. (i) Right panel: Enlarged image of (a) including the position of line grid spectra (white arrows) and single spectra locations (red/black circle). Left Panel: Single $\mathrm{d}I/\mathrm{d}U$ spectra in the star centre (black) and ring centre (red) revealing zero bias peaks of different amplitude. (j,k) Heat maps of the second derivative of the differential conductance $-\partial^2\sigma/\partial V^2$ like in Fig. 6.4 along the cross-sections marked in (i). The red/cyan line follows Δ_1/Δ_2.

area remains the same but the vortex moved and transformed from the anomalous type to the normal type. The chances of this being two entirely different vortices are rather small considering the low probability to find one in the first place. So overall, the CdGM states of individual bands seem to be almost disconnected from each other inside one vortex, which ties into the previous indications of a small interband coupling. Additionally, the vortices themselves are moveable by a change in magnetic field and can be changed from the anomalous to the normal type. One explanation for this could be, that the flux tubes are tilted with respect to the surface and that this tilting causes the displacement of the CdGM state sets by removal of the cylinder symmetry. In fact, the bright stripe (which also splits in energy) that points toward the vortex in Fig. 6.5, 6.6 and 6.7 is only present for anomalous vortices

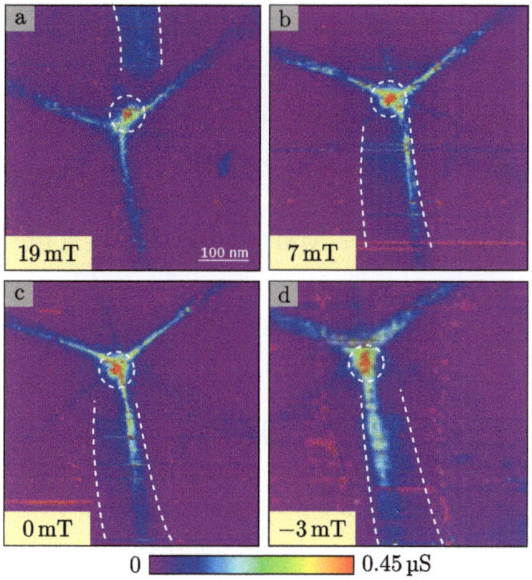

Figure 6.6: Manipulation of CdGM State Displacements: Upon slowly varying the magnetic field, the relative displacement of the star centre and ring centre (red) changes in distance and direction untied to any crystal direction. (a-d) Differential conductance at zero bias voltage around the vortex from Fig. 6.5 at decreasing magnetic fields. The position of the ring centre is marked by a white-dashed circle. The bright stripe mentioned in the main text is marked at its borders with white dashed lines

and might indicate the direction in which the magnetic field lines are tilted. For clarity, it is marked in Fig. 6.6 by white dashed lines, together with the position of the ring centre by a white dashed circle.

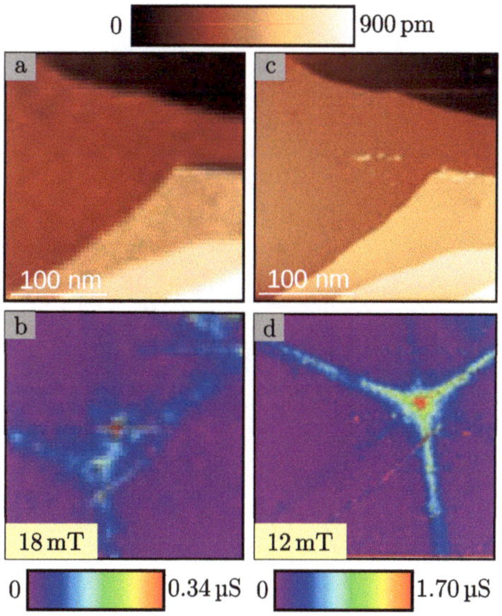

Figure 6.7: Manipulation from Anomalous to Normal Vortex: (a,c) Topographic images of the same area on the surface before (a) and after (c) the decrease in magnetic field. (b,d) Corresponding $\mathrm{d}I/\mathrm{d}U$ maps to the topographic images (a,c). After a slow decrease of the magnetic field, the anomalous vortex in (b) moved and transformed into a normal type (d). In the process, both sets of CdGM states moved.

6.3 Tunnelling into Caroli-de Gennes-Matricon States

By solving the BdG equations for a cylinder symmetric vortex line in a type-II superconductor, C. Caroli, P.G. de Gennes and J. Matricon could show that there exist fermionic bound states with energies smaller than the gap size Δ_∞ in the vortex core [229]. For $\epsilon \ll \Delta_\infty$, these Caroli-de Gennes-Matricon (CdGM) states approximately follow the linear dispersion

$$\epsilon(\mu, k_z = 0) \propto \mu \frac{\Delta_\infty}{k_F \xi} \approx \mu \frac{\Delta_\infty^2}{\epsilon_F} \tag{6.1}$$

with μ being an angular momentum quantum number that defines each state $|j\mu\rangle$ with total angular momentum j. One could think of these states as quantised Landau levels with the difference being that each of these states has to be built from the superconducting condensate and is therefore a mixture of electron and hole character. The spectrum for a single-flux vortex, calculated by Gygi and Schlüter, is shown in Fig. 6.8(a). The states are in principle discrete in energy, a zero-energy state is missing and the spectrum is inversion symmetric to the origin $\epsilon = \mu = 0$, meaning

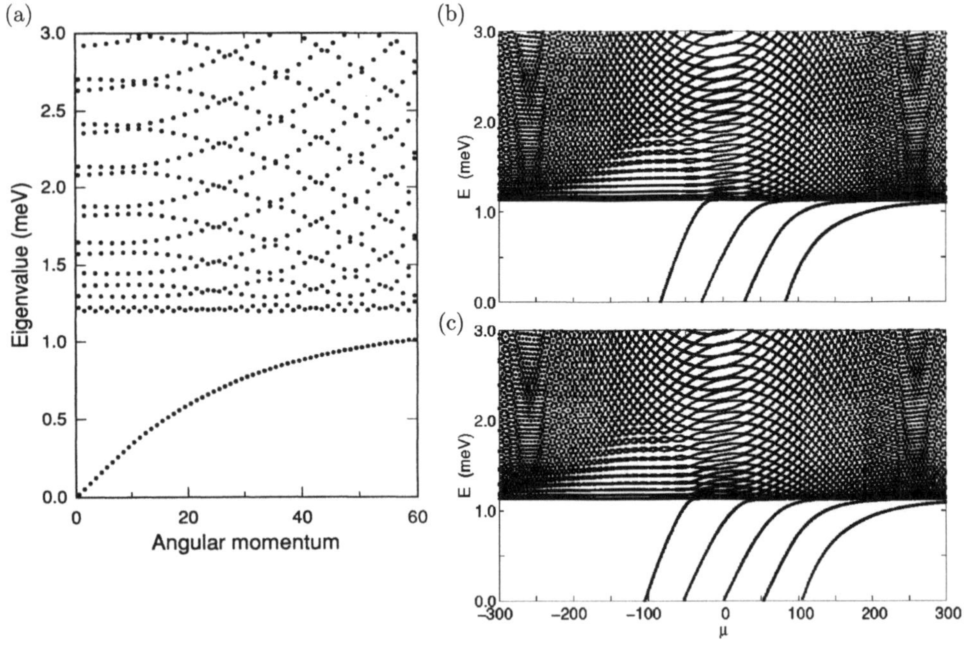

Figure 6.8: CdGM states: Quantum mechanical calculation of $\epsilon(\mu)$ for vortices of different vorticity m. (a) Energy spectrum for an $m = 1$ vortex. Reprinted figure with permission from [223]. Copyright (1991) by the American Physical Society. (b,c) Energy spectrum for an $m = 4$ (b) and $m = 5$ vortex (c). Reprinted figure with permission from [228]. Copyright (1999) by the American Physical Society.

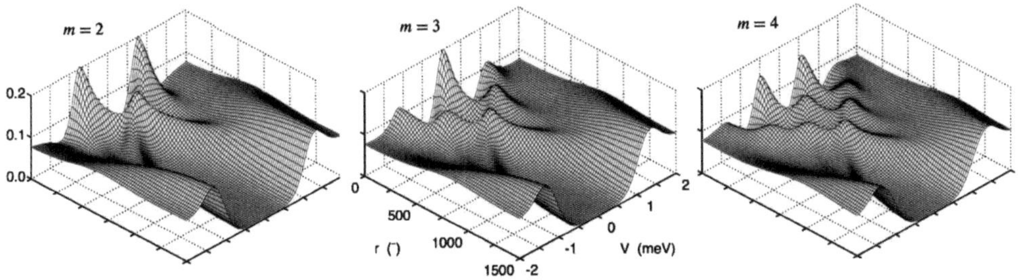

Figure 6.9: Multiquantum Vortex Fingerprint: Calculated radial conductance profile for an m-quanta vortex. Reprinted figure with permission from [228]. Copyright (1999) by the American Physical Society.

that the excitations for $\mu < 0$ have $\epsilon(\mu) < 0$. The dispersion is linear for low energy and flattens as the CdGM states approach the gap energy. While individual CdGM states can be probed on a few unconventional superconductors like $FeTe_{0.55}Se_{0.45}$ [230], that have a relatively high gap energy and small Fermi energy, for conventional superconductors the ratio Δ^2/ϵ_F is, even at ultra-cold temperatures, below the resolution limit of current techniques like STS. For Pb, this energy spacing is of the order of ~ 0.1 µeV. As a result, the states form a quasi-continuum, which is called a CdGM branch. As can be seen from comparison with results for an $m = 4$ and $m = 5$ vortex (see Fig. 6.8(b,c)) by Virtanen and Salomaa [228], there exist m individual CdGM branches in an m-quanta vortex. While all spectra are inversion symmetric to the origin, only vortices with odd winding number have a branch that crosses $E = 0$ at $\mu = 0$.

An important feature of these states, that emphasizes their similarity to Landau levels, is that the quantisation translates to some extent into real space: Gygi and Schlüter could show that a CdGM state with angular momentum μ has maximum amplitude at a distance $r \sim |\mu|/k_F$ from the vortex centre [223]. That is for a cylinder symmetric vortex and an isotropic in-plane Fermi velocity. This makes STM an excellent choice of instrument to study these states, as the whole spectrum of CdGM states can be probed in radial STS profiles of the vortex. These profiles should then exhibit the wedge shape from Fig. 6.9 with multiple dispersing branches for an m-quanta vortex. If all CdGM state branches can be disentangled and identified in these profiles, then an unambiguous determination of the vortex winding number is possible.

6.3.1 Determination of Vortex Winding Numbers

In the same year, in which Gygi and Schlüter published their results from the microscopic theory, G.E. Volovik made the connection of the low energy fermionic states inside a vortex, that he studied primarily in rotating superfluid ^{3}He-B, to topology. He found, that the number of zero energy crossings $N_{\rm zm}$, i.e. points in k-space where particle and hole spectrum meet, is a topological invariant and therefore very robust to perturbation of the system [227, 231, 232]. This means, that the overall shape of the vortex should not matter and that the number of individual CdGM state branches is simply determined by the vortex winding number m, in particular

$$N_{\rm zm} = m. \tag{6.2}$$

As discussed beforehand, in the case of the cylindrically symmetric vortex with isotropic Fermi velocity, m can easily be determined in an STM experiment. That is because the number of zero-energy peaks in a radial STS profile going from $-\xi$ to $+\xi$ through the vortex centre is $N_{\rm zm}$ and thus identical to m. The problem in deviation from the model case of perfectly cylindrical vortices and isotropic bands is that the description of the axial degree of freedom by an angular momentum is not correct, and the solutions of the problem will deviate from simple Bessel functions. Consequently, the quasiparticle spectrum, whether described fully quantum mechanically or semi-classically, has to be described by the quantum numbers (k_x, k_y, k_z). The application of the topological index theorem then proves to be more difficult from an experimental point of view, as it is unknown which line in real space crosses all diabolic points (the zero energy points in the spectrum, in this case) in k-space exactly once. Even though the states do not possess a continuous rotation symmetry anymore, the next section will show that the shape of the vortex plays an underlying role and that the discrete symmetry group of the electronic bands is inherited by the CdGM branches in real space.

6.4 Evidence of Multiflux Vortices

Apart from single-flux vortices with a diameter of $\sim 100\,\rm nm$, larger vortices could also be observed. Two examples are shown in Fig. 6.10. (a-c) show a vortex that contains two flux quanta. The zero bias conductance map (a) reveals a star shaped STM state pattern in which the arms into the $\langle 2\bar{1}\bar{1}\rangle$ direction are doubled into parallel rays. The arms into the $\langle 1\bar{2}1\rangle$ direction remain weak and the conductance in the centre of the vortex is a local minimum (see blue curve in (g)). At higher energy (b), the star arms again split into two arms each totalling four arms in each direction for this vortex. The quasi-classical calculation (a, right panel) confirms that this is the signature of a magnetic vortex with winding number $m = 2$. Again, the star shaped pattern stems from band II (top right) while band I, with a more isotropic Fermi velocity distribution, would cause a ring (bottom right). Surprisingly, the relative

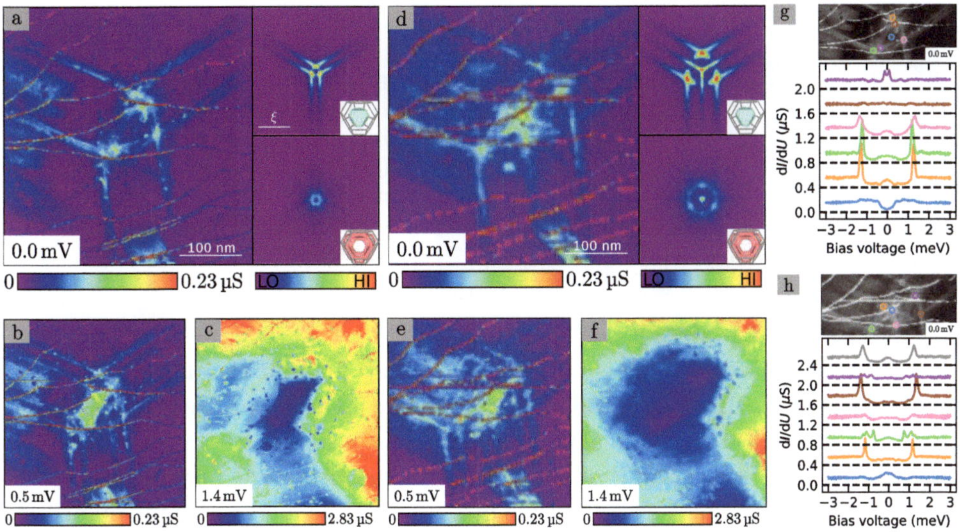

Figure 6.10: Multiflux Vortices: (a-f) dI/dU maps of an $m = 2$ (a-c) and $m = 3$ (d-f) vortex. The right panels in (a,d) display the simulated LDOS of band I (bottom) and II (top). (g,h) Single bias spectroscopies (bottom panel) at important locations of the vortex, marked with the corresponding color in the zero bias map (top panel).

magnitude of the LDOS signal from the two bands now seems to be reversed in the calculation. This is in agreement with the experimental observation, where the ring, that is predicted from band I, seems to be overshadowed by the states from band II. An analogous behaviour is found for the vortex with three flux quanta in Fig. 6.10(d-f). Here, the zero bias conductance reveals a star-shaped pattern with three arms in each $\langle 2\bar{1}\bar{1}\rangle$ direction (d) that split into six at higher energies (e). The centre of the vortex now features a local conductance maximum (blue curve in (h)). Both vortices appear larger than the single-flux vortex at $eV = \Delta_2$ (compare Fig. 6.3(h) and Fig. 6.10(c,f)), grow in lateral dimension with the number of confined flux quanta and are not round, but oval-shaped. This might once again be the result of tilted flux lines. The number of confined flux quanta is not limited to small natural numbers: In Fig. 6.11, a giant vortex with $m > 10$ is shown. It features a large but countable number of flux quanta, because the number of arms that spread from the structure are countable.

Coming back to Volovik's original topological index theorem, one can conclude that for a vortex with anisotropic Fermi velocity, it still holds that the existence/absence of a zero-energy peak in the vortex centre implies an odd/even winding number. However, it is not possible anymore to determine the exact winding number by counting the number of zero-energy crossings in a radial profile. That is because the cylinder symmetry of the problem is removed by the anisotropy in the Fermi

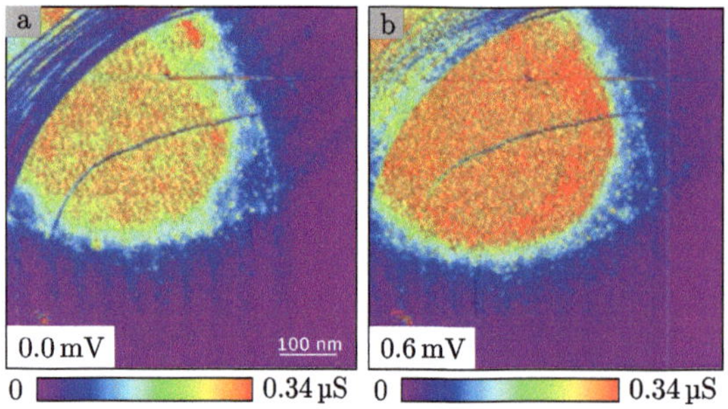

Figure 6.11: Giant Vortex: $\mathrm{d}I/\mathrm{d}U$ maps of a giant vortex with winding number $m > 10$. The STM states at zero energy (a) extend into the $\langle 11\overline{2} \rangle$ directions with more than 10 arms. At higher energies (b) the arms split like in the vortices with lower m.

velocity. As a consequence, it is not ensured that all diabolic points are crossed in the radial profile, which is crucial for the topological argument. One can, however, now reformulate the topological index theorem for a star-shaped vortex based on the experimental findings and quasi-classical calculations: The number of parallel arms l of a star in the zero-energy LDOS map seems to be identical to the number of diabolic points and thus one can once again use the topological index theorem to state that the winding number of the vortex is given by

$$m = l. \tag{6.3}$$

6.5 Vortex Interactions beyond the Ginzburg-Landau Limit

It is time to address the question, how a stabilization of single-flux and multi-flux vortices in bulk Pb is possible. With a GL parameter of $\kappa \approx 0.48$, Pb is the elemental superconductor closest to the Bogomol'nyi point, $\kappa_c = 1/\sqrt{2} \approx 0.7$, but should, according to the standard classification of superconductors, respond to an external magnetic field like a typical type-I superconductor. Apparently, the majority of domains in the intermediate state display the standard behaviour that is characteristic for type-I, i.e. large superconducting and normal regions. During the magnetic field ramp, flux flow is observed. As soon as the magnetic field is kept constant, single vortices could be stabilized, without spontaneous unpinning or drift being observed in the experiment.

This is surprising, because in superconductors of type-I, the ratio of the two characteristic length scales ξ and λ implies an attractive vortex-vortex interaction. That is because the vortex cores overlap before the screening currents see each other and two vortices merge before repulsive interactions come into play. In type-II superconductors, the contrary is the case: Here, the screening currents lead to repulsive interaction between two vortices before the vortex cores overlap. The Bogomol'nyi point with $\kappa = 1/\sqrt{2}$ manifests an infinitely degenerate point in the phase diagram where the vortex interaction potential changes sign. In an idealized superconductor, vortices should there behave like non-interacting particles [233–236]. Since the GL theory is only valid near T_c, this standard classification and argumentation is also only valid for temperatures $T \lesssim T_c$. In fact, recent works by Vagov *et al.* on the low temperature extension of the GL theory expect the degeneracy of the Bogomol'nyi point to be lifted below T_c and give rise to a phase rich transitional region in the (κ, T) plane that widens with decreasing temperature [204]. This is shown in Fig. 6.12. Of special interest is the region of multi-flux vortices right below the transitional domain with single-quantum vortices. The multi-flux vortex region seems to be narrow, but Vagov *et al.* could show in the same publication that a large Fermi velocity anisotropy in two-band superconductors can aid in widening that region further, which opens an alternative path for Pb. To believe that the vortices in this work are in a stable regime due to the temperature dependence of κ only is, however, too naive, because these theoretical calculations neglect the dependence of κ on impurities and the finiteness of sample size.

In order to test whether the above described low temperature effect is sufficient to explain the stabilization of single-flux and multi-flux vortices in bulk Pb, a control experiment at $T = 4.3\,\mathrm{K}$ was carried out. It turned out, that the stabilization of single vortices within the STM scan frame was substantially harder and only in a single case a stable vortex was found (see Fig. 6.13). This experimental observation suggests that the low temperature plays an important role in the frequent observation of single vortices, but is most likely not solely responsible. As mentioned earlier, in a

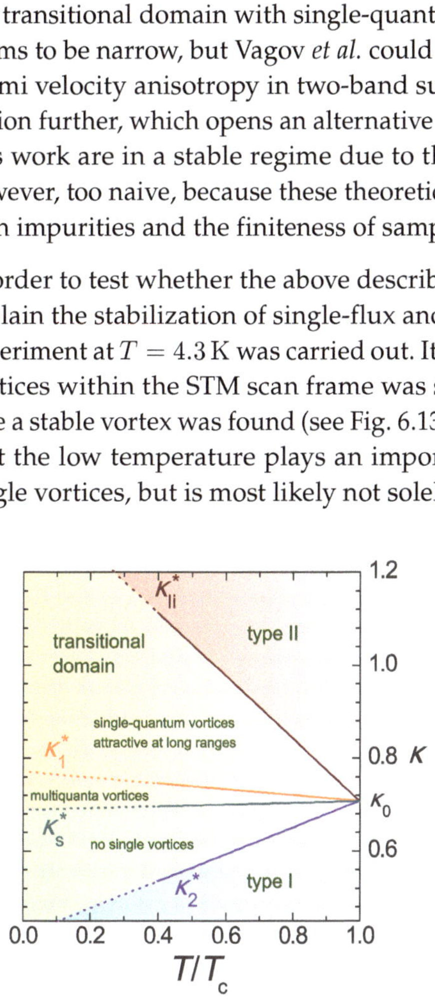

Figure 6.12: Low Temperature Transitional Region: Low temperature phase diagram of the critical GL parameter $\kappa(T)$ including the transitional region $[\kappa_2^*, \kappa_{li}^*]$ between standard types I and II. In the region $[\kappa_s^*, \kappa_1^*]$ multiquantum vortices are stable solutions. Reprinted figure with permission from [204]. Copyright (2016) by the American Physical Society.

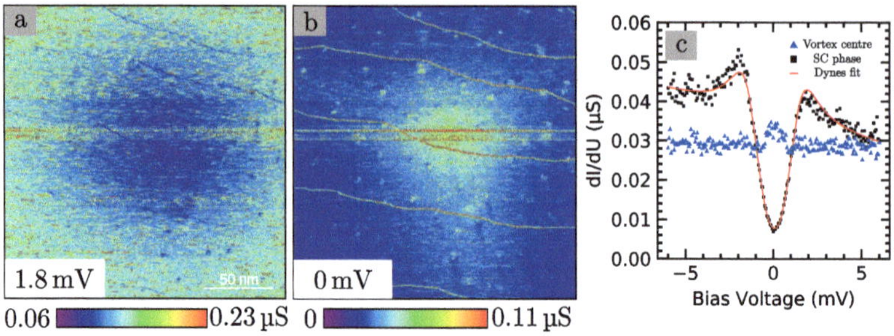

Figure 6.13: Vortex at $T \lesssim T_c$: (a,b) $\mathrm{d}I/\mathrm{d}U$ maps of a magnetic vortex at $T = 4.3\,\mathrm{K}$ and $B = 0\,\mathrm{mT}$. (c) Bias spectroscopies far away from the vortex (black squares) and in the centre of the vortex (blue triangles) in (a). The red line shows a fit of Dynes form to the differential conductance in the superconducting region with a gap size of $\Delta = 1.24\,\mathrm{meV}$.

real single crystal of Pb, the GL parameter is also affected by crystal defects, which can at the same time lead to flux pinning as discussed in Chap. 2.1.2.

Extended defects with a size $d \sim \xi$ are effective pinning centres for flux. An example for such a defect is a screw dislocation, as shown in Fig. 6.14(a). The strong pinning energy scales linearly with the volume $V_c = \xi^2 L$, in which the screw dislocation runs through the vortex line, and the condensation energy of Cooper pairs $E_{\mathrm{cond}} = \mu_0 H_c(0)^2/2$ [237]. The ends of screw dislocations are clearly visible in STM topographies, but no trend of flux pinning at these sites on the surface was obvious during the experiment. However, at stacking fault tetrahedrons, extended defects that are unique in fcc metals, vortices were regularly pinned. Due to their intricate structure and electronic footprint, a detailed analysis can be found in Chap. 7. In a collective pinning mechanism, defects with a size $d \ll \xi$ can also effectively trap flux albeit with a smaller scaling ratio. As an example, the weak collective pinning by point defects is depicted in Fig. 6.14(b). In the potential of randomly distributed point defects, the pinning energy scales with $\sqrt{V_c}$. To some extent, the vortex line can sacrifice elastic energy in order to run through parts of the crystal with higher defect concentrations. The length scale L_c on which the vortex line remains relatively straight is therefore a measure for the pinning strength [237].

All in all, it seems unlikely that the isolated vortices in this work are a result of a high κ alone. In that case, one would expect either a lot more vortices ($\kappa_1^* < \kappa < \kappa_{li}^*$) or domains of all kinds of sizes ($\kappa_2^* < \kappa < \kappa_1^*$). Instead, one mostly finds isolated vortices coexisting with µm-large domains. The low temperature does play an important role, as the control experiment at 4.3 K demonstrated. It prevents thermally driven flux creep and puts bulk Pb farther away from the standard type-I regime in the intermediate state so that attractive vortex-vortex interactions are smaller. As a result, lower pinning energies are needed to trap flux on a defect. One can see

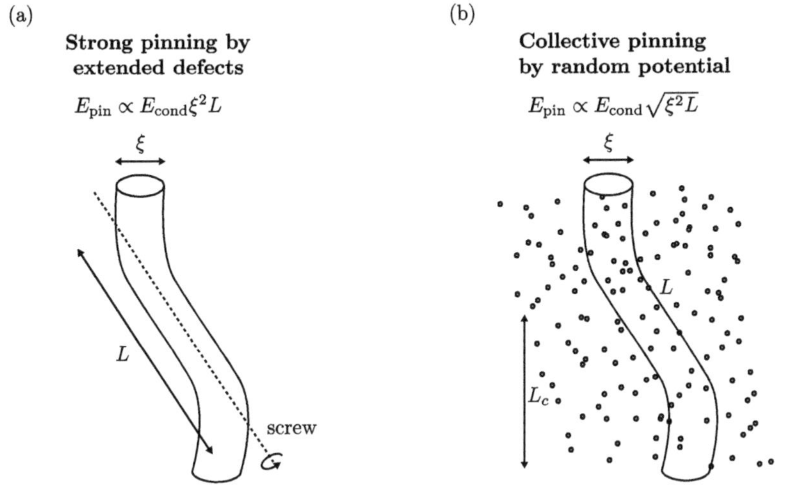

Figure 6.14: Vortex Pinning Mechanisms: (a) Strong pinning by an extended defect ($d_{\text{def}} > \xi$), like a screw dislocation: The pinning energy is proportional to the volume, in which the screw dislocation runs parallel through the vortex line. (b) Weak pinning by point defects ($d_{\text{def}} \ll \xi$): Small defects can collectively pin flux. The elastic vortex line bends to run through the crystal volume with the most defects. The length scale on which the vortex line remains straight L_c is a measure for the strength of this pinning mechanism. The collective pinning energy only scales like the square root of the total flux line volume. Adapted from Ref. [237].

indications for strong pinning on stacking fault tetrahedrons but not on single screw dislocations. Additionally, impurities can aid the larger defects by a collective pinning mechanism.

6.6 Summary

Although bulk lead (Pb) is classified as a prototypical type-I superconductor, it was demonstrated that the intermediate state hosts single and multi-flux quanta vortices at mK temperatures. As stability criteria for the existence of the vortices, the influence of temperature and defect density on the effective GL parameter κ were discussed. By probing the quasiparticle LDOS inside the vortices with STM and comparing their signature with simulations, it was possible to identify the CdGM states of the two superconducting bands of Pb. The spatial LDOS pattern created by the CdGM states provides a robust method to determine the vorticity or winding number. The robustness is rooted in the fact that the number of diabolic points in the CdGM state spectrum are a topological invariant of the vortex. Simulations of the single and multiflux quanta vortices in the quasi-classical Eilenberger theory, in conjunction with DFT calculated band structure of Pb, support the experimental findings and interpretations. Tilted vortices were useful to demonstrate a low interband coupling for the two superconducting bands, as their respective CdGM state patterns were independently moveable. Additionally, they allowed for a definite assignment of the two gaps $\Delta = (1.26 \pm 0.02)\,\mathrm{meV}$ and $\Delta = (1.40 \pm 0.02)\,\mathrm{meV}$ to the two superconducting bands of Pb.

7 Quasiparticle and Flux Trapping in Topological Defects of a Two-band Superconductor

Despite being topological defects themselves, vortices in dimension $d = 3$ are not expected to host topologically non-trivial bound states inside their core [238]. The situation changes when topological crystal defects, i.e. extended defects with dimension $D \geq 1$ like dislocations ($D = 1$) or twin boundaries ($D = 2$), are added to the mix [239]. In this chapter, the focus lies on stacking fault tetrahedra (SFTs) in Pb(111). These extended defects, that are unique to fcc metals, are efficient pinning centres for flux in the intermediate state, as shown in the differential conductance map in Fig. 7.1. After elucidating their formation and structural properties, their impact on the low energy electronic spectrum is highlighted. Lastly, the interaction of these crystal defects with the CdGM states of a pinned vortex are discussed. This chapter is the logical continuation of Chap. 6 toward the combination of extended

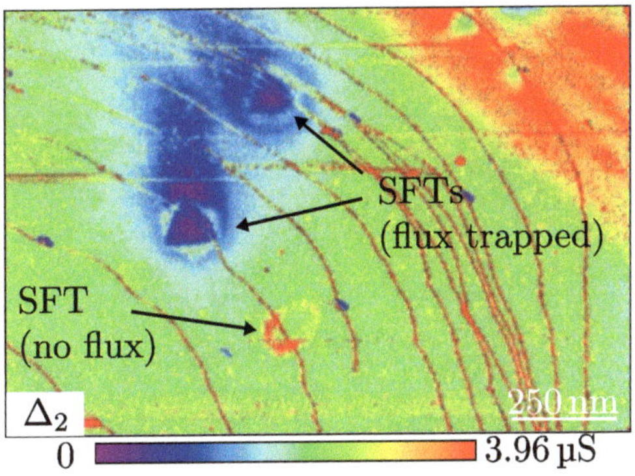

Figure 7.1: Flux Traps: Differential conductance map on intermediate state Pb(111) at $eU = \Delta_2 = 1.4\,\text{mV}$. Areas of low conductance (purple/blue) mark normal conducting regions and areas of high conductance (green/yellow/red) mark superconducting regions. The triangles with a surrounding hexagon of higher conductance are the electronic footprints of SFTs. In two out of three visible SFTs magnetic flux is trapped.

crystal defects and (multiflux) vortices and therefore builds up on findings from the vortex study in Pb.

7.1 Stacking Fault Tetrahedra

In face-centred cubic crystals, the preferred glide plane for dislocations is the (111) plane and the intersections of all available {111} planes form a regular tetrahedron. Therefore, the movement of dislocations is most conveniently explained at the Thompson tetrahedron (TT). As shown in Fig. 7.2, the Burgers vector of a (partial) dislocation can be easily drawn in the TT by the connection of corner points (Latin letters) or face centres (Greek letters).

Perfect dislocations (edge or screw type) are found at the sides of the TT, so the notation for its Burgers vector is the Latin letter for the beginning and end corner point, e.g. BA, which corresponds to the bulk direction $\frac{1}{2}[1\bar{1}0]$ in Miller index notation. The unfolded TT in Fig. 7.2(b) serves as a dictionary to translate between Miller index and TT notation. Under applied stress, the glide of a perfect dislocation along a {111} plane is hindered by the hexagonal atomic configuration in the plane and it is energetically favorable for the atoms to move in zig-zag motion via the hcp

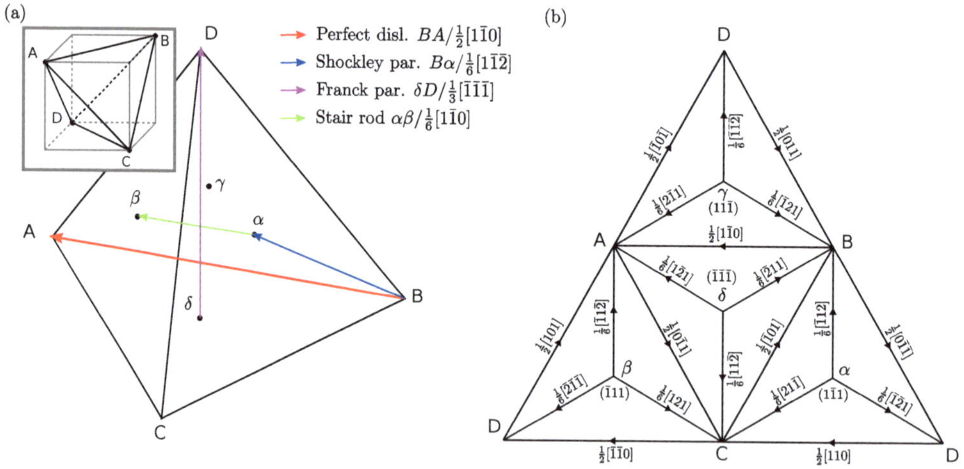

Figure 7.2: Thompson Tetrahedron: (a) The {111} glide planes in an fcc crystal form a regular tetrahedron - the Thompson tetrahedron (TT). The Burgers vector of (partial) dislocation lines are all found as connections between corner points or face centres of the tetrahedron. (b) The unfolded TT with marked bulk crystal directions serves as a dictionary to translate between Miller indices and TT notation, e.g. $BA \rightarrow \frac{1}{2}[1\bar{1}0]$.

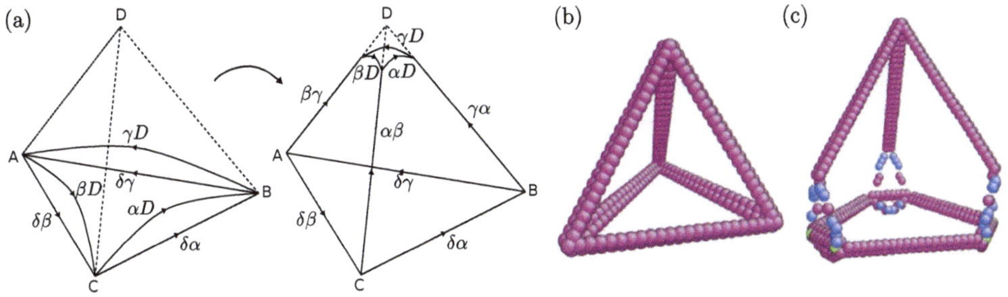

Figure 7.3: Silcox-Hirsch Mechanism: (a) The SFT of a frustrated configuration of stair-rod dislocations is built from a triangular Franck partial loop following the dislocation reactions (7.2) and (7.3). Adapted from [240, p. 99]. (b,c) Molecular dynamics (MD) simulations demonstrate that the Silcox-Hirsch mechanism can create an SFT from a triangular (b) or scalene hexagonal platelet of vacancies (c). Reprinted from [241], Copyright (2007), with permission from Elsevier.

hollow site on the (111) plane. Formally, the perfect dislocation dissociates into a pair of *Shockley partial* dislocations:

$$BA \to B\gamma + \gamma A. \tag{7.1}$$

If the stacking fault energy is low enough, the dislocation can travel through the crystal as an extended dislocation, in which a leading Shockley partial and a trailing Shockley partial enclose an intrinsic stacking fault. This means that between the two Shockley partials the stacking sequence in $[111]$ direction is not $ABCAB$.. but $ABABC$...

The stacking sequence inside an fcc crystal can also be faulted by the removal or insertion of one close-packed $\{111\}$ layer of atoms with finite lateral dimension (island). The boundary line of the stacking fault forms a closed loop and is called a *Franck partial* dislocation. The difference to the Shockley partial is that its Burgers vector points perpendicular to the $\{111\}$ plane, which is faulted, e.g. $\delta D \equiv \frac{1}{3}[\bar{1}\bar{1}\bar{1}]$. The Shockley partial and the perfect dislocation, whose Burgers vector lie within a $\{111\}$ plane, are so-called *glissile* dislocations, because they can freely move in the glide plane under applied stress. The Franck partial on the other hand is a *sessile* dislocation and can only move by climb [240, p. 93].

The SFT is a nanocrystal in the shape of a regular or truncated tetrahedron with stacking faults on each of its $\{111\}$ facets and commonly observed in metals of low stacking-fault energy, like Cu, Ag or Au [242–244]. It is believed to form in the *Silcox-Hirsch* mechanism [240, p. 98-99]: First, vacancies or impurities condense to a triangular platelet, say in the ABC plane, then the bounding Franck partial dissociates on three adjacent $\{111\}$ facets into a so-called *stair-rod* partial (two Greek letters) and a Shockley partial dislocation (one Greek, one Latin letter):

$$\begin{aligned} \delta D &\rightarrow \delta\alpha + \alpha D \quad (\text{on BCD}) \\ \delta D &\rightarrow \delta\beta + \beta D \quad (\text{on ACD}) \\ \delta D &\rightarrow \delta\gamma + \gamma D \quad (\text{on ABD}). \end{aligned} \tag{7.2}$$

The *stair-rod* partials remain in a locked configuration along the triangle edges (on ABC) and repel the glissile Shockley partials due to their parallel Burgers vectors. The Shockley partials then move up the tetrahedron faces and pull up additional stair-rod dislocations on the intersection edges (AD, BD, CD), as shown in Fig. 7.3(a), by the dislocation reactions

$$\begin{aligned} \alpha D + D\beta &\rightarrow \alpha\beta \quad (\text{along CD}) \\ \beta D + D\gamma &\rightarrow \beta\gamma \quad (\text{along AD}) \\ \gamma D + D\alpha &\rightarrow \gamma\alpha \quad (\text{along BD}). \end{aligned} \tag{7.3}$$

If the stair-rod dislocations culminate in the tip of the TT (D), then one ends up with a regular SFT, if they stop their movement before, one ends up with a truncated SFT. Atomistic simulations could show that the Silcox-Hirsch mechanism can, under certain circumstances, also develop SFTs from a scalene hexagonal platelet of vacancies [241], which explains the shape of SFTs, observed on Au(111) [244]. The results of these simulations for a triangle and scalene hexagon as the starting vacancy plate are also shown in Fig. 7.3(b,c).

7.1.1 Stacking Fault Tetrahedra in Pb(111)

In the experiments from Chap. 6 and 7, the extended defects which are able to trap flux are interpreted as SFTs based on their size, alignment with the glide planes and electronic footprint. They are most likely created during the hot sputtering process although the dependence of abundance and shape of them on temperature, ion flux and ion energy has not been studied in this work. Figure 7.4(a-d) shows the topographies of four different exemplary SFTs in Pb(111). The white dotted lines correspond to contours of the LDOS as seen in the respective low bias differential conductance maps (e-h) in the superconducting state of the crystal. In (a) and (b) the SFT lies at the surface, which is visible as a scalene hexagonal depression in the topographic image. The apparent negative height is a fraction of one single (111) lattice plane distance. The white contour lines in the $[2\bar{1}\bar{1}]$ and equivalent crystal directions on the surface are in fact the directions in which one would expect projections of the stair-rod dislocations $\alpha\beta$, $\gamma\alpha$ and $\beta\gamma$ (cf. Fig. 7.3(a)). The difference in length of these contour lines between (a) and (b) hints towards different evolutionary stages, so different truncation levels, of the SFTs.

By that logic, the SFT in (a) has developed less of a tetrahedron before being truncated by the surface, while the SFT in (b) has developed further, with longer stair-rods. Of course, this is a highly hypothetical interpretation, because the bulk structure is only inferred from a small amount of structural information and the electronic DOS at the surface. Given the fact, that the surface energies in Pb favour vacancy or impurity islands of scalene hexagonal shape in the {111} planes [245], a formation of SFTs like in Fig. 7.3(c) seems most probable. In the topographic images in Fig. 7.4(c,d) a depression like in (a,b) is missing. The SFTs seem to be buried beneath the surface, yet close enough for the LDOS at the surface to confirm their existence. In (c), the SFT seems to bind a non-threading screw dislocation of which both ends are visible at the surface. In (d), no influence on the surface structure by the sub-surface SFT is visible. Unlike the surfacing SFTs, these two SFTs have a rather similar area ratio between the inner scalene hexagon (almost triangle) and the outer hexagon.

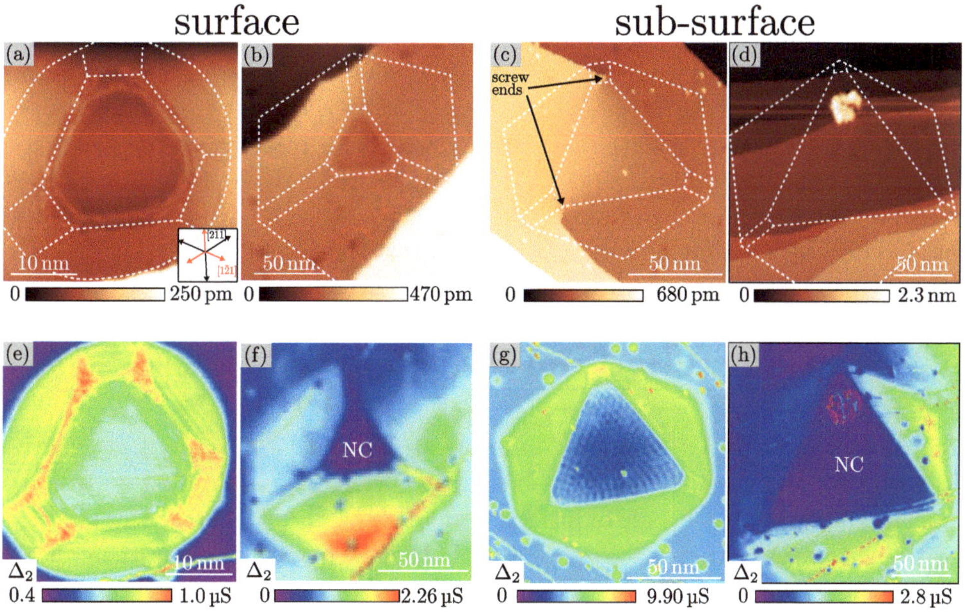

Figure 7.4: Appearance of SFTs: (a-d) Topographic images of areas containing (truncated) SFTs. The white lines outline where the respective map of differential conductance exhibits clear contours of the LDOS. Surfacing SFTs (a,b) are visible by a hexagonal shaped depression of a fraction of a lattice plane distance. Sub-surface SFTs (c,d) are only detectable through their electronic footprint. (e-f) dI/dU-maps at $U = 1.4\,\text{meV}$ displaying the corresponding LDOS to the topographies in (a-d). In (f) and (h) flux is threaded through the centre of the defect rendering it normal conducting.

It is possible that these truncated SFTs represent the equilibrium shape acquired universally in the bulk, but a definite answer lacks statistical evidence.

In $\mathrm{d}I/\mathrm{d}U$ maps at $eU = \Delta_2$ (Fig. 7.4(e-h)) (superconducting state) surface and sub-surface SFTs are easily recognizable by a hexagonal region of $\approx 30 - 150\,\mathrm{nm}$ (depending on SFT size) of increased conductance with a cutout, that is usually a scalene hexagon (or almost a triangle), of lower conductance. Rod-like contours of the LDOS can be seen along the sides of the inner triangle/scalene hexagon and along the $\langle 1\bar{2}1\rangle$ directions inside the outer hexagon. The former lie within the $\{111\}$ planes, in which the stacking faults are expected and the latter fall together with the projected directions in which the stair-rod dislocations are expected. In Fig. 7.4 (f,h), flux is threaded through the SFT rendering its inner structure normal conducting (dark purple). The general contour lines of the LDOS, however, remain intact. In Fig. 7.4(e), a standing wave pattern in the green hexagonal (almost round) part of the LDOS is visible, which is absent in the normal conducting state (see Fig. E.1 in Appendix E). In Fig. 7.4(g), this standing wave pattern is either absent or could not be resolved. Here, however, a very clear quasiparticle interference (QPI) pattern is obvious in the inner triangle. Similar particle-in-the-box states have also been found by means of STM in the SFTs of Au(111) and Ag(111) [243, 244] and nanocavities in Pb(111) [245, 246]. The energies at which these states are observed are slightly larger in the SFTs of Au and Ag ($E \approx 1.6 - 4.2\,\mathrm{meV}$) and much larger for the nanocavities ($E \approx 400\,\mathrm{meV}$), which is reasonable considering the size of the respective confinement box. Since stacking-faults are known to have tremendous effect on the electronic band structure, the following section is dedicated to the origin of the QPI pattern and the LDOS contrast around an SFT.

7.2 Band Filter Effect and Quantum Well States

Before turning to the QPI patterns, it makes sense to take a closer look at why the LDOS at the gap edge Δ_2 seems to exhibit such strong local contrast with sharp transitions on the length scale of just a few atoms. As it turns out, this is not limited to the energy Δ_2. At $eU = \Delta_1$ the outer hexagon appears lower in differential conductance than the cut-out or the superconducting surrounding but still with the same sharp contour, as Fig. 7.5(a) shows for the SFT from Fig. 7.4(c,g). The source of this contrast reversal in the outer hexagon becomes clear when one looks at the single $\mathrm{d}I/\mathrm{d}U$ spectra in Fig. 7.5(b) along the line indicated in (a): Inside the inner triangle (points 0-6), the coherence peak at Δ_2 appears to be missing, inside the outer hexagon (points 8-13), the tunnelling amplitude into the coherence peak of Δ_2 exceeds the one for Δ_1 and outside the hexagon (points 15-25), the spectra regain the characteristic coherence peak heights for the Pb(111) surface. This behaviour is best seen in a plot of the fitted amplitude σ_i of the BCS curve for the gap Δ_i as a function of spectroscopic location, like in Fig. 7.5(c). In the bottom panel, the ratio $A = \frac{\sigma_2}{\sigma_1}$ and the asymmetry $\chi = \frac{\sigma_1 - \sigma_2}{\sigma_1 + \sigma_2}$ are plotted. While the tunnelling amplitude into the two coherence peaks of Pb was demonstrated to be dependent on the tunnelling direction with respect to the crystal orientation [39], here, only vertical tunnelling into (111) oriented planes is expected above the SFT[1]. Together with the fact that the superconducting properties cannot change on a length scale much

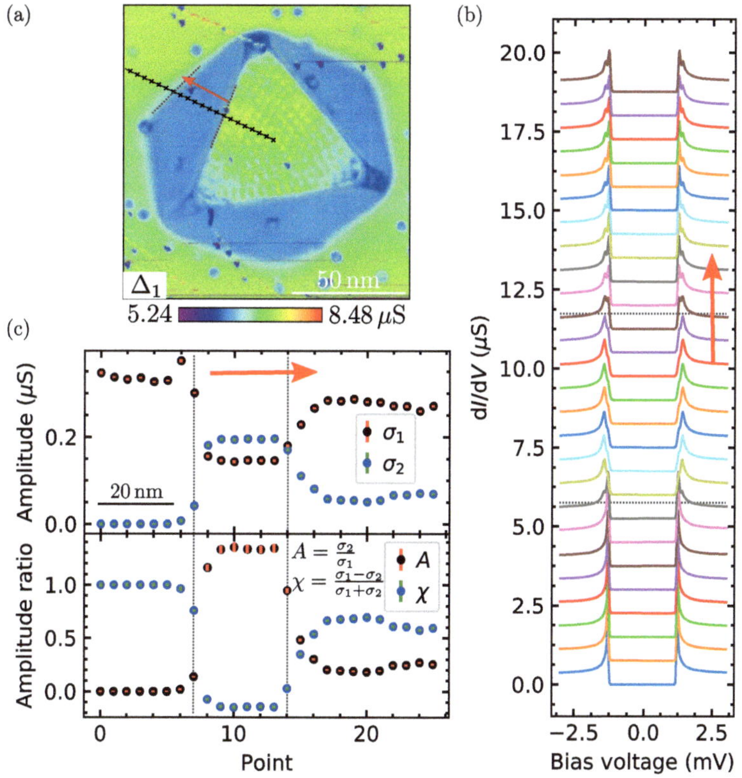

Figure 7.5: Band Filter Effect: (a) dI/dU map ($eU = \Delta_1$) of SFT from Fig.7.4(c,d). Black crosses mark single spectra location and the red arrow indicates the direction in which the points were sequentially probed. (b) Two-gap BCS fits to the spectra obtained along the line marked in (a). The single spectra are offset in y-direction for visibility. (c) Top panel: Scaling factor (amplitude) σ_1/σ_2 of the BCS curves with Δ_1/Δ_2 along the line marked in (a). Bottom panel: Amplitude ratio $A = \sigma_2/\sigma_1$ and asymmetry $\chi = (\sigma_1 - \sigma_2)/(\sigma_1 + \sigma_2)$ along the same line.

[1] Even in the presence of screw dislocations, the distorted planes around the dislocation retain the character of (111) planes because there is no stacking fault.

shorter than the coherence length, it is most likely that the seen effect is caused by a constraint imposed on the movement of normal conducting electrons in the system. It seems that the stacking faults of the SFT act as a band filtering energy barrier that traps QPs of band I with the energy Δ_1 inside the tetrahedron and hinders QPs of band II with energy Δ_2 from entering it. As a consequence, a QPI pattern of QPs from band I[2] is formed inside the tetrahedron (see Fig. 7.5(a)) and standing waves of QPs from band II are formed outside the tetrahedron (see Fig. 7.4(e)) as a result of their backscattering. What the authors of Ref. [247] claim to be "enhanced superconductivity" around nanocavities in Pb(111) could also be explained by the band selective scattering of QPs with energy Δ_1 and Δ_2, which would not be resolved in their experiment due to insufficient energy resolution. What is puzzling, is that the normal conducting state, so far, showed no signs of this band filter effect.

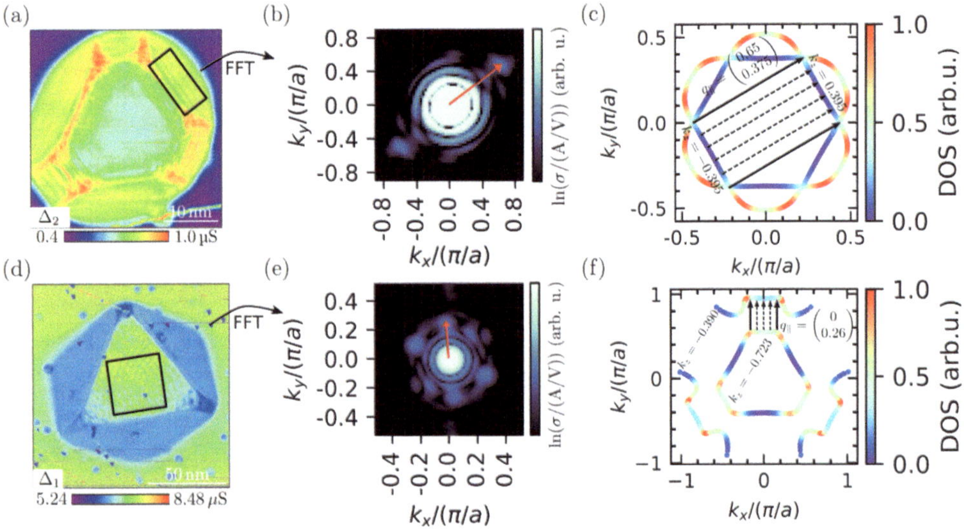

Figure 7.6: Scattering Vectors: (a) dI/dU map ($eU = \Delta_2$) of the SFT from Fig. 7.4(a) reveals standing wave fronts in the green region oriented toward the $\langle 2\bar{1}\bar{1}\rangle$ and the $\langle 1\bar{2}1\rangle$ directions. (b) FFT of the region marked in (a). The scattering vector (red arrow) has the length $|\boldsymbol{q}_{\|}| = 0.78\,\pi/a$ in the $[2\bar{1}\bar{1}]$ bulk ([11] surface) direction. (c) Corresponding nesting vector $\boldsymbol{q}_{\|} = (0.65, 0.375)\,\pi/a$ on the Fermi surface of band II responsible for the scattering vector in (b). (d) dI/dU map ($eU = \Delta_1$) of the SFT from Fig. 7.4(c) reveals QPI pattern inside the inner triangle. (e) Analogous to (b): the scattering vector is $|\boldsymbol{q}_{\|}| = 0.26\,\pi/a$ in the $[11\bar{2}]$ direction. (f) Analogous to (c): The corresponding nesting vector on the FS of band I has $\boldsymbol{q}_{\perp} \approx 0.333\,\pi/a$ and $\boldsymbol{q}_{\|} = (0, 0.26)\,\pi/a$. In (c,f) k_x and k_y point in the $[1\bar{1}0]$ and $[11\bar{2}]$ direction. The axes in (b,e) are slightly misaligned to these directions (see real space directions in Fig. 7.4(a)).

[2] The QPI pattern seen in the same defect at $eU = \Delta_2$ in Fig. 7.4(g) is caused by the oscillation of the coherence peak amplitude at Δ_1 and the sensitivity of the measurement to its high energy flank at Δ_2.

In a local STS measurement, one gains exceptional real space resolution for the sacrifice of momentum information, because the pure eigenstates in real space are the Fourier transformed eigenstates in k-space. Techniques like angle-resolved photoemission spectroscopy (ARPES) in turn probe the pure momentum eigenstates that contain no local information. In the presence of crystal impurities, however, these two pictures are mixed because the translation invariance is broken. Especially in inhomogeneous superconductors like Bi2212 [248, 249], researchers developed the technique of Fourier-transform STM, where the probed LDOS of the QPs could be linked to scattering vectors on the Fermi surface (FS) and ultimately explain discrepancies between STM and ARPES studies. The coherent scattering on a defect leads to oscillations of the LDOS in real space with the period $\lambda = 2\pi/|\boldsymbol{q}|$, where $\boldsymbol{q} = \boldsymbol{k} - \boldsymbol{k}'$ is the scattering vector.

The Fast-Fourier transform[3] of the QPI pattern marked in Fig. 7.6(a) in the outer hexagon reveals the scattering vector with length $|\boldsymbol{q}_{\|}| = 0.78\,\pi/a$ in the $[2\bar{1}\bar{1}]$ direction (see (b)). This vector connects opposite sides of the Fermi surface of band II (along $U-\Gamma-U$ and $K-\Gamma-K$) where the DOS is relatively small but this is overcompensated by the connection of very large portions of the FS by this single vector, as shown in (c). The same technique can be applied to the QPI pattern in Fig. 7.6(d) at energy Δ_1. One finds the scattering vector with $|\boldsymbol{q}_{\|}| = 0.26\,\pi/a$ in all $\langle 2\bar{1}\bar{1}\rangle$ and $\langle 1\bar{2}1\rangle$ directions in the (111) plane, as shown in (e). The corresponding 3D scattering vector with the same $\boldsymbol{q}_{\|}$, shown in (f), connects opposite sides of the smaller necks in the FS of band I (along $U - X - U$) in the $\Gamma - X$ direction. Also here, the connected FS parts are almost parallel to each other and have a relatively high DOS. Because the QPI pattern in (d) is limited to a closed box (the inside of the SFT), one needs to be a bit more careful in the interpretation. That is because the boundaries of the box impose another (larger) periodicity on the electronic wavefunctions on top of the lattice periodic crystal potential.

The Schrödinger equation of a free particle in an equilateral triangle (2D) box has an analytical solution [250, 251] that has been experimentally demonstrated by the authors of Ref. [243] in the STM study of SFTs on Ag(111). Ag(111) has a surface state of quadratic dispersion and SFTs with a triangular ground plane making it an optimal system to study the QWSs of the triangular 2D box. The electrons in this surface state are a quasi-free 2D electron gas with constant DOS so that the interference pattern inside the SFT is solely determined by the box geometry and energy. On Pb(111), such a surface state is absent, because band II does not have a bulk band gap in this direction. In nanocavities of Pb(111) [245, 246], the QWSs are still dictated by the box geometry because the box size is of the same order as the

[3] A Hanning window function was used in the Fast-Fourier transforms.

Figure 7.7: Free particle in an Equilateral Triangle Box: Analytical solution for the LDOS of a free electron with quantum numbers $(p,q) = (\frac{2}{3}, 17\frac{2}{3})$ in an equilateral triangle following Ref. [250].

lattice constant and therefore overshadows the bandcharacter of the electrons. The QPI in an SFT of superconducting Pb(111) like in Fig. 7.6(a) should in principle be viewed as the result of Bogolyubov QPs just above the gap energy that inherit the Bloch character of electrons[4] at the Fermi surface and scatter at the stacking faults of the tetrahedron. Therefore, they should scatter predominantly with scattering vectors that connect parallel parts of the FS with high DOS and indeed this is what was seen in Fig.7.6.(d-f). Still, the interference pattern shows striking resemblance to the QWS obtained in the model of a free electron gas inside a triangular 2D box, that is displayed in Fig. 7.7. Considering the fact that the QWSs are not (or not clearly) visible in the defect from Fig. 7.6(a), it seems that in Fig. 7.6(d) the SFT/triangular box has just the right size to support QWSs with a periodicity in real space that corresponds to strong scattering vectors.

In order to shed more light on the nature of the interference pattern, it should first be made sure that the same pattern is seen at opposite bias voltage which is to be expected for the near-gap QPs of a superconductor. Secondly, the evolution of the interference pattern with energy could be measured in order to learn more about the dispersion of the QPs. And lastly, a larger number of SFTs should be investigated to make concluding remarks on the influence of the box size and shape.

[4] In this case, the box size is much larger than the periodicity of the lattice potential making a description in the basis of Bloch states reasonable.

7.3 Electronic Edge State

Stacking faults have already proved to be efficient strong pinning centres for flux, at least in high-temperature superconductors with columnar defects [252, 253]. Because a tendency toward flux pinning at SFTs was seen in this work, it was possible to pin a single-flux vortex inside an SFT.

In Fig. 7.8(a), the zero-bias conductance map around the defect from Fig. 7.4(c) features the star-shaped CdGM states characteristic for a single-flux vortex. The central line of the vortex (centre of CdGM state pattern) lies inside the SFT as a comparison to the $\mathrm{d}I/\mathrm{d}U$ map at $eU = \Delta_2$ (e) clarifies. The star arms in the $\langle 2\bar{1}\bar{1}\rangle$ directions, which stem from band II, are shorter than before. The states from band I, so the weak arms in $\langle 1\bar{2}1\rangle$ directions and the higher energy CdGM states (b-c), that produce a ring in the real space pattern, do not seem to cross the triangular outline of the SFT. A logical explanation would be that this effect is also caused by the band filtering of the stacking faults. Although not well resolved, in (b,c) the high-energy CdGM states seem to produce a QPI pattern similar to the one observed for the gap edge QPs. One additional feature is striking in the map in (a): In the presence of the pinned single-flux vortex, an additional state, which is localized on the edge of the

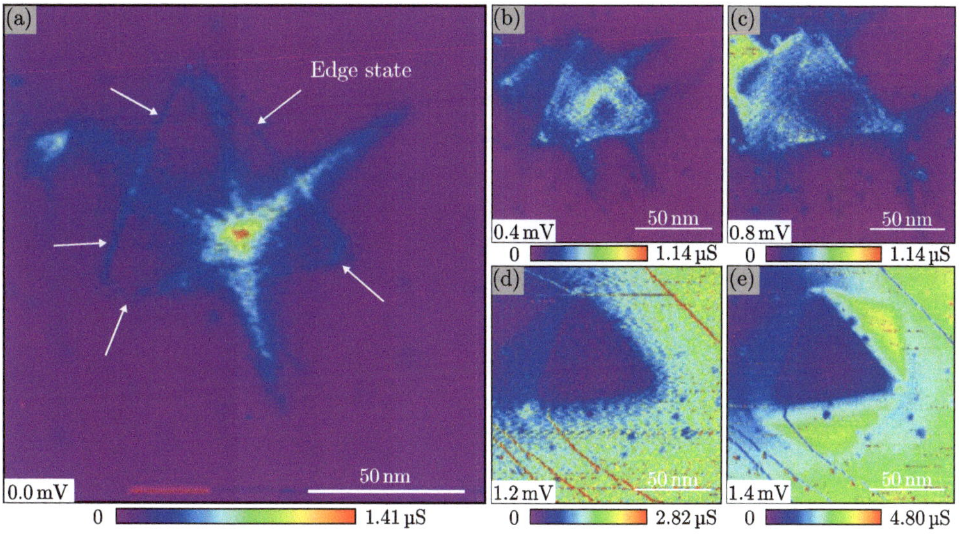

Figure 7.8: Single-flux Vortex in SFT: $\mathrm{d}I/\mathrm{d}U$ maps of the SFT from Fig. 7.4(c) at $B = 6\,\mathrm{mT}$: a single-flux vortex is threading the tetrahedron. (a) At $U = 0.0\,\mathrm{mV}$ the CdGM state pattern of the vortex is less extended (shorter star arms) and along the SFT edge a zero-energy edge state emerged. (b-e) At $U = 0.4$, 0.8, 1.2, $1.4\,\mathrm{mV}$, the regular behaviour of CdGM states in Pb is seen. The CdGM states also seem to produce a QPI pattern inside the tetrahedron.

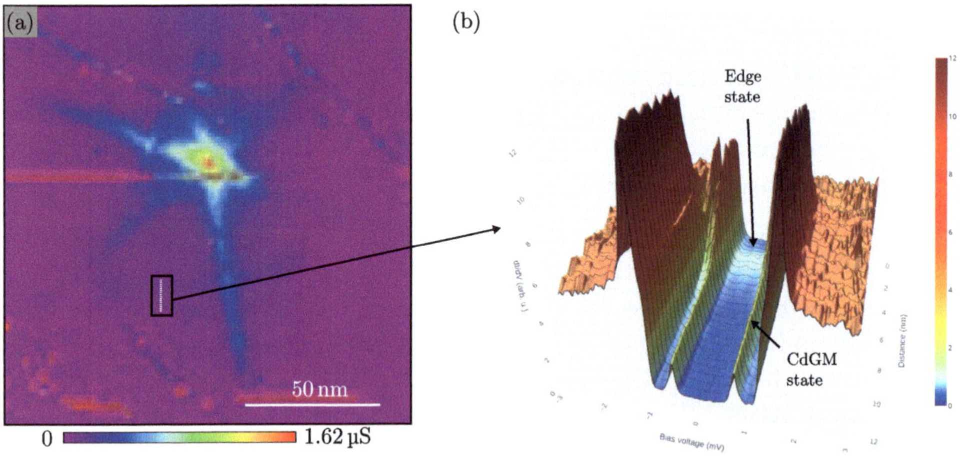

Figure 7.9: Edge State: (a) dI/dU map of same area as in Fig. 7.8 after decreasing the magnetic field to $B = -2\,\mathrm{mT}$. The vortex moved further up but remains inside the SFT and the edge state prevails. (b) Line profile of the differential conductance as function of bias voltage along the white line marked in (a). The CdGM states do not cross zero energy. At the position of the SFT edge a very localized peak centred around zero energy emerges.

SFT, appeared. This state is absent in the purely superconducting phase (see Fig. E.2 in Appendix E).

Even after slightly shifting the vortex line by a reduction of the applied magnetic field from $B = 6\,\mathrm{mT}$ to $-2\,\mathrm{mT}$, the edge state persists, as demonstrated in Fig. 7.9(a). A closer look at the differential conductance spectrum along the white dotted line (b) makes it clear that the edge state is distinct from the CdGM branch in the spectrum. The latter does not move considerably in energy along the line profile, the former appears exactly at the triangular contour of the SFT where the stacking fault line is assumed to be.

In order to continue the discussion of inherent differences between the CdGM states and the edge state on the basis of physical quantities, the localisation and "lifetimes"[5] or spectral bandwidths of these states were compared. This is shown in Fig. 7.10.

[5] The spectral bandwidth of CdGM states around zero energy will for reasons of readability be associated with a "lifetime", although, this interpretation is strictly not allowed. In a measurement at a single tip position, several low energy CdGM states are expected to overlap. STS can never distinguish between single life-time broadened states and bands if the energetic spacing of states is below the resolution capability of the instrument. The general conditions for each single measurement are, however, the same, so the deduced spectral widths will still only differ if the properties of the states themselves are different.

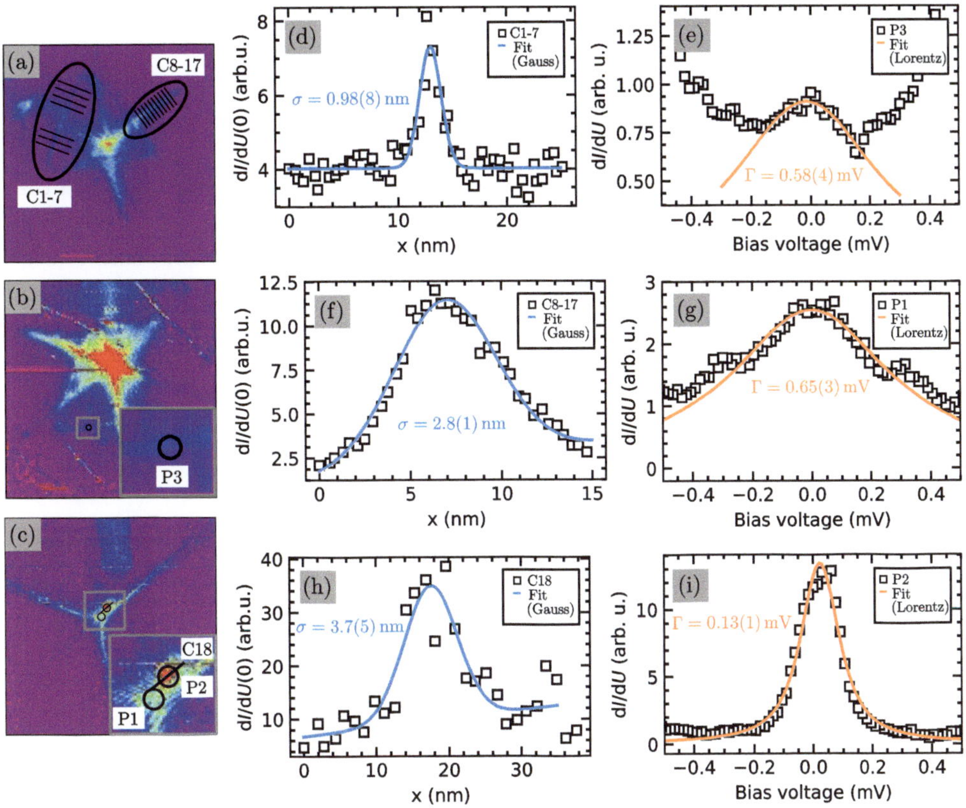

Figure 7.10: Localisation and Lifetimes: (a-c) Zero bias dI/dU maps of normal single flux vortices trapped on an SFT (a,b) (vortex from Fig. 7.8, 7.9) and of an anomalous vortex (b) (vortex from Fig. 6.5) with marked cross sections C1-18 (lines) and point spectroscopies P1-3 (circles). The localisation of the zero energy edge state (d), the zero energy CdGM state of band II (f) and the zero energy CdGM state of band I (h) was determined through a fit of the average cross section $dI/dU(U=0,\mathrm{x})$ to a Gaussian. The corresponding lifetimes were determined by a fit of the point bias spectroscopies to a Lorentzian (e,g,i).

In the normal vortex pinned on an SFT (a) the localisation of the edge state can be extracted from cross sections of the dI/dU signal at $U = 0\,\mathrm{mV}$ across the edge. To enhance the signal-to-noise ratio, several cross-sections (C1-7) were averaged. This averaged line profile was fitted to a generalized Gaussian of the form

$$f(x) = \frac{C}{\sqrt{2\pi}\sigma}\exp\left(-\frac{(x-x_0)^2}{2\sigma^2}\right) + b + cx \tag{7.4}$$

with the characteristic width σ. The lifetime $\tau = 1/\Gamma$ was determined from a single point spectroscopy on the edge (P3) shown in the pinned vortex in (b). The $dI/dU(U)$

curve was reduced by a constant background and then the part around $U \approx 0$ was fitted to a Lorentzian of the form

$$g(U) = \int_{-\infty}^{\infty} \mathrm{d}U' \frac{C}{(U' - U_0)^2 + (\Gamma/2)^2} \frac{\partial n_F(U' - U)}{\partial U} \tag{7.5}$$

with the characteristic scattering rate $\Gamma = 1/\tau$. For the CdGM states, the fact that the states from the different bands can be probed independently, especially in an anomalous vortex (c), was used. The localisation and lifetime of the zero energy CdGM states of band II (star pattern) were extracted from the cross-sections C8-17 (f) and the point spectroscopy P1 (g). The results for the zero energy CdGM states of band I (point at P2) are shown in (h) and (i). The respective localisation widths σ and scattering rates Γ from the fits are indicated in the figure. Even from this small set of data, significant differences between edge state and the CdGM states

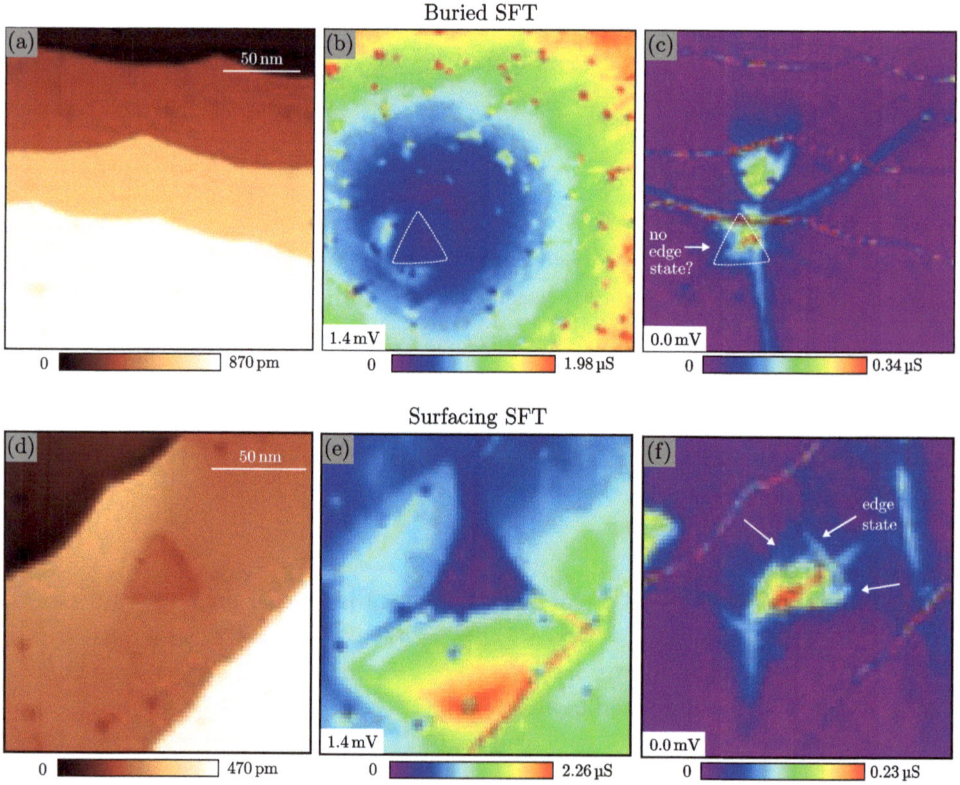

Figure 7.11: Edge State in Buried and Surfacing SFTs: (a-c) Buried SFT with a pinned single-flux vortex showing no sign of an edge state at the triangular contour of the SFT (white dotted line). (d-f) Surfacing SFT with a pinned single-flux vortex showing a clear edge state at zero bias. (a,d) Topographic images of the area containing the SFT. (b,e) $\mathrm{d}I/\mathrm{d}U$ maps at $U = 1.4\,\mathrm{mV}$ of the flux-penetrated SFT. (c,f) Corresponding $\mathrm{d}I/\mathrm{d}U$ maps at $U = 0\,\mathrm{mV}$.

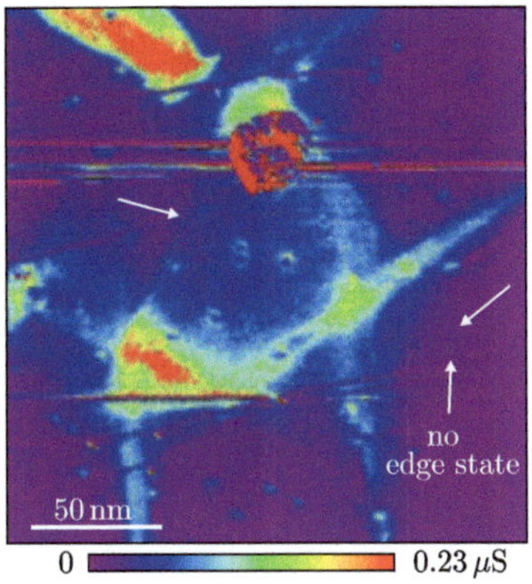

Figure 7.12: $m = 2$ **Vortex in SFT**: $\mathrm{d}I/\mathrm{d}U$ map at $U = 0.0\,\mathrm{mV}$ of an $m = 2$ vortex trapped inside the sub-surface SFT from Fig. 7.4(d). No clear edge state is visible.

become apparent. The edge state has a much smaller lifetime than the zero energy CdGM states of band I, it is more comparable to the zero energy CdGM states of band II. It is, however, much more localized than the CdGM states of both bands as a comparison of the deduced localisation widths σ shows. From this analysis, a clear one-to-one correspondence between the edge state and the CdGM states is not seen. The fact that a sudden change from normal conducting region to superconducting region is seen at the edge of this SFT is intriguing in the sense that this edge state may play an important role for the flux screening.

Comparing two SFTs with a pinned single-flux vortex, one buried beneath the surface (Fig. 7.11(a-c)) and one at the surface (Fig. 7.11(d-f)), indicates that the edge state (marked with white arrows) is only visible for surfacing SFTs. This would not be too surprising, considering the fact that the edge state is also strongly localized in the lateral direction. Hence, it might be present but not detectable on the surface for buried SFTs. Of course, for a convincing argument a higher number of SFTs with pinned vortices would have to be analysed.

The pinning of an $m = 2$ vortex inside the defect from Fig. 7.4(d) has also been achieved. In the zero-bias conductance map in Fig. 7.12, the characteristic footprint of the $m = 2$ vortex is seen, yet no edge state is visible, where the SFT edge is seen in the map at $eU = \Delta_2$ (cf. Fig. 7.4(h)). Whether the invisibility of the edge state in this case is due to the even vorticity of the vortex, the fact that this defect is buried deeper beneath the surface or has an entirely different reason, cannot be concluded from the available data.

While a topological origin of this state is far from certain, none of the here reported properties are in conflict with it. In fact, some of them constitute necessary conditions for a topological edge state. In summary, the following properties of the SFT edge state were found:

- It is not seen if the whole area containing the SFT is superconducting or normal conducting.
- It is only present, if a vortex is pinned on the SFT.
- Its energetic position is the Fermi energy.
- An abrupt change from the normal conducting region inside the SFT to the superconducting region outside the SFT is seen on the atomic length scale ($a \ll \xi$), when the vortex is pinned and the edge state is present.
- The spectral width and localisation of it are different from the CdGM states crossing the Fermi energy. Especially the localisation is much stronger. Perpendicular to the edge, it is only spread across a few atoms.
- The position falls together with a region of large disturbance to the electronic band structure, the stacking faults of the SFT.
- An even/odd effect for the vorticity is seen (although the vorticity does not have to be the cause, as mentioned above). Several single flux ($m = 1$) vortices pinned on an SFT appear together with the edge state. The $m = 2$ vortex pinned on an SFT appears without it.

7.4 Summary

By hot sputtering a Pb(111) single crystal with keV argon ions and subsequent annealing, SFTs were formed. In the superconducting state, these extended crystal defects have tremendous influence on the LDOS of QPs. They are responsible for a memorable LDOS pattern at the gap energy that cannot be explained by structural deformations alone. A large change in the electronic band structure is expected at the stacking faults that surround the tetrahedral nanocrystal. This boundary seems to act as a band selective QP filter. It traps QPs of one band inside the SFT in the form of quantum well states (QWSs) and prevents the QPs of the other band from entering. Since analogous measurements in the normal state do not show these effects, the SFT may potentially constitute a Cooper pair box. Upon pinning vortices on SFTs, the transition between normal conducting region and superconducting region is in certain cases reduced to a length scale much lower than the known bulk coherence length. Concurrently with this, a zero-energy edge state emerges at the boundary of the SFT. A series of systematic tests with the purpose of fathoming the origin of this state could not reject a topological cause.

8 Conclusion and Outlook

In the scope of this thesis, the full potential of a mK scanning tunnelling microscope (STM) was used to study three very different types of superconducting materials that are constant subject of fundamental research: A granular superconductor, a conventional multi-band superconductor and unconventional cuprate superconductors. In the presence of localized spins, vortices or crystal defects the elastic part of the tunnelling current revealed a large variety of interesting quasiparticle (QP) effects that include Yu-Shiba-Rusinov (YSR) states, Caroli-de Gennes-Matricon (CdGM) states and interband couplings. Outside of the superconducting gap, inelastic tunnelling events contributed significantly to the tunnelling current. A special focus was given to antiferromagnetic spin fluctuations and potential plasmon or Higgs modes near a superconductor-to-insulator transition (SIT). In the analysis part of the thesis, the deeper understanding of CdGM states and observed inelastic excitations required quasiclassical calculations in the framework of 3D Eilenberger equations and an advanced deconvolution algorithm that were independently programmed by the author.

In quest of the bosonic glue in cuprate superconductors, we demonstrated how the effective Eliashberg function can be extracted from IETS data in the superconducting state. The extracted boson spectral function of these d-wave superconductors is in good agreement with the expected integrated spin spectrum, which validates the spin-fermion model. As momentum information is usually lost in a tunnelling experiment, an essential ingredient in the success of this method was the dominant influence of hot-spots on the Fermi surface on the superconducting density of states. The real space imaging power of the STM could be used further to examine the position dependence of the bosonic excitation and potentially even perform QPI measurements at the right energies. The minimum energy of the inelastic scattering channel is expected to vary locally due to the inhomogeneity of Δ.

We put together the first STM study of granular aluminium (grAl) (that we know of) and published the results in Physical Review B [146]. The systematic study of samples with different oxygen content could confirm many microscopic properties of these superconducting films that were previously derived from bulk measurements, like electronic transport or optical spectroscopy. These include a homogeneous superconducting gap, as well as a gap enhancement and grain decoupling that scale with the oxygen concentration. In addition, on individual grains in the high-resistivity film, the tunnelling electrons could easily excite a long-lived bosonic mode. It is likely that this mode corresponds to phase (or amplitude) fluctuations of the superconducting gap near the SIT. Perhaps the most important result in the high-resistivity grAl film was the observation of YSR states. The careful analysis of position dependent tunnelling spectra hints towards localized spins that reside in the oxide barriers between the aluminium grains. If this is indeed the case, then this has important implications for many Al/AlO_x based superconducting devices as spin scattering can potentially limit their performance. In principle, these localized spins should also appear in the natural oxide of aluminium. First attempts to find YSR states in the natural oxide of a polycrystalline aluminium film could, however, not confirm them. The measurements should be repeated on a thicker oxide before final conclusions can be drawn. As the tunnelling barrier increases with a thicker surface oxide layer, different experimental techniques may be required.

It is also worthwhile to revisit supposedly well understood conventional superconductors like lead (Pb) with state-of-the-art STM setups as ultra-low temperatures and better energy resolution can provide new insights into the microscopic understanding of the superconducting properties. In this work, it was demonstrated that the intermediate state of a bulk Pb single crystal can naturally host stable vortices with single and multiple flux quanta at mK temperatures. We thus have to rethink our traditional classification of superconductors into type-I and type-II on the microscopic scale and at temperatures far below the critical temperature. Using the exceptional energy resolution of a 25 mK STM, it was possible to study the interplay of CdGM states from two different bands inside a single vortex line. Additionally, the long theoretically predicted LDOS patterns of CdGM states in multiflux vortices could for the first time be confirmed experimentally and their immediate connection to the vortex winding number was demonstrated. While the CdGM states of a vortex in an s-wave superconductor are not considered to be topologically protected bound states, topologically non-trivial boundary states can in principle exist at topological defects. A combination of the multiflux vortices with extended crystal defects seemed like the obvious next step.

We saw that stacking fault tetrahedra (SFTs) have a detrimental effect on the local quasiparticle DOS of Pb. Apparently, the weak interband coupling of the two superconducting bands leads to band selective scattering processes at the SFT boundary and confinement effects. Stacking faults are known to have a relatively large effect on the electronic band structure of the normal state, yet it remains an open question which role superconductivity plays in this case. The SFTs can also act as pinning centres for multiflux vortices. As it turns out, the CdGM states are also affected by the SFT boundary and in certain cases, a pinned single-flux vortex lead to a zero-energy edge state localized on the SFT boundary. Whether this state is topological in the quasiparticle sense remains to be seen, but the present findings do not yield a categorical argument against it.

Superconducting phenomena often span very different classes of superconductors. One example is the SIT that is of central importance in various high temperature superconductors (HTSs), but also in granular superconductors. Another example is multiband superconductivity which is widely accepted to be present in iron-based superconductors and the elementary superconductor Pb but has been mostly overlooked up until the discovery of two-band superconductivity in MgB_2 in 2001. This thesis shows that it is worthwhile to revisit superconductors that have been around for several decades and that we can, when equipped with state-of-the-art experimental and theoretical techniques, reveal previously hidden properties. When being open for surprises, we can often find them in places where we did not expect them. Of special noteworthiness are the findings of localized spins in grAl with important implications for the use of Al/AlO_x based superconducting devices and the existence of stable (multiflux) vortices in bulk Pb, which provide optimal conditions to study vortex core states in combination with multiband effects. Especially with the discovery of an intriguing edge state, that emerged where vortex states were perturbed by extended crystal defects, we hope to initiate further research towards the realization of a Majorana zero mode (MZM).

Bibliography

[1] Nishijima, S., Eckroad, S., Marian, A., Choi, K., Kim, W. S., Terai, M., Deng, Z., Zheng, J., Wang, J., Umemoto, K., Du, J., Febvre, P., Keenan, S., Mukhanov, O., Cooley, L. D., Foley, C. P., Hassenzahl, W. V. & Izumi, M.: *Superconductivity and the environment: a Roadmap*. Supercond. Sci. Technol. **26**, 113001. doi:10.1088/0953-2048/26/11/113001 (2013) (cit. on p. 1).

[2] Onnes, H.: *The resistance of gold and pure mercury at helium temperatures*. Commun. Phys. Lab. Univ. Leiden **120b** (1911) (cit. on p. 1).

[3] Onnes, H.: *The disappearance of the resistance of mercury*. Commun. Phys. Lab. Univ. Leiden **122b** (1911) (cit. on p. 1).

[4] Onnes, H.: *On the sudden change in the rate at which the resistance of mercury disappears*. Commun. Phys. Lab. Univ. Leiden **124c** (1911) (cit. on p. 1).

[5] Matthias, B. T.: *Chapter V Superconductivity in the Periodic System* (ed Gorter, C. J.) 138–150. doi:10.1016/S0079-6417(08)60104-3 (Elsevier, 1957) (cit. on p. 1).

[6] Pickett, W. E.: *The other high-temperature superconductors*. Physica B: Condensed Matter. *Proceedings of the Symposium on Wave Propagation and Electronic Structure in Disordered Systems* **296**, 112–119. doi:10.1016/S0921-4526(00)00787-0 (2001) (cit. on p. 1).

[7] Bednorz, J. G. & Müller, K. A.: *Possible high T_c superconductivity in the Ba−La−Cu−O system*. Z. Physik B - Condensed Matter **64**, 189–193. doi:10.1007/BF01303701 (1986) (cit. on pp. 1, 51).

[8] Schilling, A., Cantoni, M., Guo, J. D. & Ott, H. R.: *Superconductivity above 130 K in the Hg–Ba–Ca–Cu–O system*. Nature **363**, 56–58. doi:10.1038/363056a0 (1993) (cit. on p. 1).

[9] Drozdov, A. P., Eremets, M. I., Troyan, I. A., Ksenofontov, V. & Shylin, S. I.: *Conventional superconductivity at 203 kelvin at high pressures in the sulfur hydride system*. Nature **525**, 73–76. doi:10.1038/nature14964 (2015) (cit. on p. 2).

[10] Drozdov, A. P., Kong, P. P., Minkov, V. S., Besedin, S. P., Kuzovnikov, M. A., Mozaffari, S., Balicas, L., Balakirev, F., Graf, D., Prakapenka, V. B., Greenberg, E., Knyazev, D. A., Tkacz, M. & Eremets, M. I.: *Superconductivity at 250 K in lanthanum hydride under high pressures*. Nature **569**, 528–531. doi:10.1038/s41586-019-1201-8 (2019) (cit. on p. 2).

[11] Ashcroft, N. W.: *Metallic Hydrogen: A High-Temperature Superconductor?* Phys. Rev. Lett. **21**, 1748–1749. doi:10.1103/PhysRevLett.21.1748 (1968) (cit. on p. 2).

[12] Maxwell, E.: *Isotope Effect in the Superconductivity of Mercury*. Phys. Rev. **78**, 477–477. doi:10.1103/PhysRev.78.477 (1950) (cit. on p. 2).

[13] Fröhlich, H.: *Theory of the Superconducting State. I. The Ground State at the Absolute Zero of Temperature*. Phys. Rev. **79**, 845–856. doi:10.1103/PhysRev.79.845 (1950) (cit. on p. 2).

[14] Fröhlich, H.: *Interaction of electrons with lattice vibrations*. Proceedings of the Royal Society of London. Series A. Mathematical and Physical Sciences **215**, 291–298. doi:10.1098/rspa.1952.0212 (1952) (cit. on p. 2).

[15] Cooper, L. N.: *Bound Electron Pairs in a Degenerate Fermi Gas*. Phys. Rev. **104**, 1189–1190. doi:10.1103/PhysRev.104.1189 (1956) (cit. on p. 2).

[16] Bardeen, J., Cooper, L. N. & Schrieffer, J. R.: *Theory of Superconductivity*. Phys. Rev. **108**, 1175–1204. doi:10.1103/PhysRev.108.1175 (1957) (cit. on pp. 2, 12).

[17] Eliashberg, G. M.: *Interactions between electrons and lattice vibrations in a superconductor*. Sov. Phys. JETP **11**, 696–702 (1960) (cit. on pp. 2, 12, 17).

[18] McMillan, W. L.: *Transition Temperature of Strong-Coupled Superconductors*. Phys. Rev. **167**, 331–344. doi:10.1103/PhysRev.167.331 (1968) (cit. on pp. 2, 18).

[19] Giaever, I.: *Energy Gap in Superconductors Measured by Electron Tunneling*. Phys. Rev. Lett. **5**, 147–148. doi:10.1103/PhysRevLett.5.147 (1960) (cit. on pp. 2, 31).

[20] Abeles, B., Cohen, R. W. & Cullen, G. W.: *Enhancement of Superconductivity in Metal Films*. Phys. Rev. Lett. **17**, 632–634. doi:10.1103/PhysRevLett.17.632 (1966) (cit. on pp. 3, 70).

[21] Abanov, A., Chubukov, A. V. & Schmalian, J.: *Fingerprints of spin mediated pairing in cuprates*. Journal of Electron Spectroscopy and Related Phenomena **117-118**, 129–151. doi:10.1016/S0368-2048(01)00251-1 (2001) (cit. on pp. 3, 30).

[22] Abanov, A., Chubukov, A. V. & Schmalian, J.: *Quantum-critical theory of the spin-fermion model and its application to cuprates: Normal state analysis*. Advances in Physics **52**, 119–218. doi:10.1080/0001873021000057123 (2003) (cit. on pp. 3, 28, 29, 53, 55).

[23] Zasadzinski, J. F., Ozyuzer, L., Miyakawa, N., Gray, K. E., Hinks, D. G. & Kendziora, C.: *Correlation of Tunneling Spectra in* $Bi_2Sr_2CaCu_2O_{8+\delta}$ *with the Resonance Spin Excitation*. Phys. Rev. Lett. **87**, 067005. doi:10.1103/PhysRevLett.87.067005 (2001) (cit. on p. 3).

[24] Zasadzinski, J. F., Coffey, L., Romano, P. & Yusof, Z.: *Tunneling spectroscopy of* $Bi_2Sr_2CaCu_2O_{8+\delta}$ *: Eliashberg analysis of the spectral dip feature*. Phys. Rev. B **68**, 180504. doi:10.1103/PhysRevB.68.180504 (2003) (cit. on p. 3).

[25] Zasadzinski, J. F., Ozyuzer, L., Coffey, L., Gray, K. E., Hinks, D. G. & Kendziora, C.: *Persistence of Strong Electron Coupling to a Narrow Boson Spectrum in Overdoped* Bi_2 Sr_2 $CaCu_2$ $O_{8+\delta}$ *Tunneling Data*. Phys. Rev. Lett. **96**, 017004. doi:10.1103/PhysRevLett.96.017004 (2006) (cit. on p. 3).

[26] Fong, H. F., Bourges, P., Sidis, Y., Regnault, L. P., Ivanov, A., Gu, G. D., Koshizuka, N. & Keimer, B.: *Neutron Scattering from Magnetic Excitations in* $Bi_2Sr_2Ca_2O_{8+\delta}$. Nature **398**, 588–591. doi:10.1038/19255 (1999) (cit. on pp. 3, 54, 66).

[27] Jandke, J., Yang, F., Hlobil, P., Engelhardt, T., Rau, D., Zakeri, K., Gao, C., Schmalian, J. & Wulfhekel, W.: *Unconventional pairing in single FeSe layers*. Phys. Rev. B **100**, 020503. doi:10.1103/PhysRevB.100.020503 (2019) (cit. on pp. 3, 58, 62, 67).

[28] Hlobil, P., Jandke, J., Wulfhekel, W. & Schmalian, J.: *Tracing the Electronic Pairing Glue in Unconventional Superconductors via Inelastic Scanning Tunneling Spectroscopy*. Phys. Rev. Lett. **118**, 167001. doi:10.1103/PhysRevLett.118.167001 (2017) (cit. on pp. 3, 30, 55, 57, 62).

[29] Grünhaupt, L., Spiecker, M., Gusenkova, D., Maleeva, N., Skacel, S. T., Takmakov, I., Valenti, F., Winkel, P., Rotzinger, H., Wernsdorfer, W., Ustinov, A. V. & Pop, I. M.: *Granular aluminium as a superconducting material for high-impedance quantum circuits*. Nat. Mater. **18**, 816–819. doi:10.1038/s41563-019-0350-3 (2019) (cit. on pp. 3, 69, 73).

[30] Grünhaupt, L.: *Granular aluminium superinductors* ISBN: 9783731509561. doi:10.5445/KSP/1000097320 (KIT Scientific Publishing, Karlsruhe, 2019) (cit. on pp. 3, 69, 89).

[31] Wang, J. I.-J. & Oliver, W. D.: *An aluminium superinductor*. Nat. Mater. **18**, 775–776. doi:10.1038/s41563-019-0401-9 (2019) (cit. on pp. 3, 69).

[32] Winkel, P., Borisov, K., Grünhaupt, L., Rieger, D., Spiecker, M., Valenti, F., Ustinov, A. V., Wernsdorfer, W. & Pop, I. M.: *Implementation of a Transmon Qubit Using Superconducting Granular Aluminum*. Phys. Rev. X **10**, 031032. doi:10.1103/PhysRevX.10.031032 (2020) (cit. on pp. 3, 69, 78).

[33] Rieger, D., Günzler, S., Spiecker, M., Paluch, P., Winkel, P., Hahn, L., Hohmann, J. K., Bacher, A., Wernsdorfer, W. & Pop, I. M.: *Granular aluminium nanojunction fluxonium qubit*. Nat. Mater. 1–6. doi:10.1038/s41563-022-01417-9 (2022) (cit. on pp. 3, 69).

[34] Koch, R. H., DiVincenzo, D. P. & Clarke, J.: *Model for $1/f$ Flux Noise in SQUIDs and Qubits*. Phys. Rev. Lett. **98**, 267003. doi:10.1103/PhysRevLett.98.267003 (2007) (cit. on pp. 3, 69).

[35] Bachar, N., Lerer, S., Hacohen-Gourgy, S., Almog, B. & Deutscher, G.: *Kondo-like behavior near the metal-to-insulator transition of nanoscale granular aluminum*. Phys. Rev. B **87**, 214512. doi:10.1103/PhysRevB.87.214512 (2013) (cit. on pp. 3, 69, 79, 80).

[36] Nagamatsu, J., Nakagawa, N., Muranaka, T., Zenitani, Y. & Akimitsu, J.: *Superconductivity at 39 K in magnesium diboride*. Nature **410**, 63–64. doi:10.1038/35065039 (2001) (cit. on pp. 3, 19).

[37] Bouquet, F., Fisher, R. A., Phillips, N. E., Hinks, D. G. & Jorgensen, J. D.: *Specific Heat of Mg 11 B 2 : Evidence for a Second Energy Gap*. Phys. Rev. Lett. **87**, 047001. doi:10.1103/PhysRevLett.87.047001 (2001) (cit. on pp. 3, 19).

[38] Iavarone, M., Karapetrov, G., Koshelev, A. E., Kwok, W. K., Crabtree, G. W., Hinks, D. G., Kang, W. N., Choi, E.-M., Kim, H. J., Kim, H.-J. & Lee, S. I.: *Two-Band Superconductivity in MgB_2*. Phys. Rev. Lett. **89**, 187002. doi:10.1103/PhysRevLett.89.187002 (2002) (cit. on pp. 3, 19, 20).

[39] Ruby, M., Heinrich, B. W., Pascual, J. I. & Franke, K. J.: *Experimental Demonstration of a Two-Band Superconducting State for Lead Using Scanning Tunneling Spectroscopy*. Phys. Rev. Lett. **114**, 157001. doi:10.1103/PhysRevLett.114.157001 (2015) (cit. on pp. 3, 19, 20, 97, 98, 123).

[40] Moshchalkov, V., Menghini, M., Nishio, T., Chen, Q. H., Silhanek, A. V., Dao, V. H., Chibotaru, L. F., Zhigadlo, N. D. & Karpinski, J.: *Type-1.5 Superconductivity*. Phys. Rev. Lett. **102**, 117001. doi:10.1103/PhysRevLett.102.117001 (2009) (cit. on p. 3).

[41] Poole, C. K., Farach, H. & Creswick, R.: *Handbook of superconductivity* 1st ed. (ed Poole, C. P.) ISBN: 0125614608 (Academic Press, San Diego [u.a.], 2000) (cit. on pp. 3, 97, 100).

[42] Cren, T., Fokin, D., Debontridder, F., Dubost, V. & Roditchev, D.: *Ultimate Vortex Confinement Studied by Scanning Tunneling Spectroscopy*. Physical Review Letters **102**. doi:10.1103/PhysRevLett.102.127005 (2009) (cit. on p. 4).

[43] Ge, J., Gutierrez, J., Cuppens, J. & Moshchalkov, V. V.: *Observation of single flux quantum vortices in the intermediate state of a type-I superconducting film*. Phys. Rev. B **88**, 174503. doi:10.1103/PhysRevB.88.174503 (2013) (cit. on p. 4).

[44] Boersch, H., Kunze, U., Lischke, B. & Rodewald, W.: *Observation of the mixed state in films of type I superconductors (Pb)*. Physics Letters A **44**, 273–274. doi:10.1016/0375-9601(73)90917-1 (1973) (cit. on p. 4).

[45] Ning, Y. X., Song, C. L., Wang, Y. L., Chen, X., Jia, J. F., Xue, Q. K. & Ma, X. C.: *Vortex properties of two-dimensional superconducting Pb films*. J. Phys.: Condens. Matter **22**, 065701. doi:10.1088/0953-8984/22/6/065701 (2010) (cit. on p. 4).

[46] Tinkham, M.: *Introduction to superconductivity* 2. ed. ISBN: 9780486435039 (Dover Publ., Mineola, NY, 2004) (cit. on pp. 9, 88).

[47] Bardeen, J. & Stephen, M. J.: *Theory of the Motion of Vortices in Superconductors*. Phys. Rev. **140**, A1197–A1207. doi:10.1103/PhysRev.140.A1197 (1965) (cit. on p. 9).

[48] Nambu, Y.: *Quasi-Particles and Gauge Invariance in the Theory of Superconductivity*. Phys. Rev. **117**, 648–663. doi:10.1103/PhysRev.117.648 (1960) (cit. on p. 10).

[49] Goldstone, J.: *Field theories with « Superconductor » solutions*. Nuovo Cim **19**, 154–164. doi:10.1007/BF02812722 (1961) (cit. on p. 10).

[50] Goldstone, J., Salam, A. & Weinberg, S.: *Broken Symmetries*. Phys. Rev. **127**, 965–970. doi:10.1103/PhysRev.127.965 (1962) (cit. on p. 10).

[51] Shimano, R. & Tsuji, N.: *Higgs Mode in Superconductors*. Annu. Rev. Condens. Matter Phys. **11**, 103–124. doi:10.1146/annurev-conmatphys-031119-050813 (2020) (cit. on p. 10).

[52] Greiter, M.: *Is electromagnetic gauge invariance spontaneously violated in superconductors?* Annals of Physics **319**, 217–249. doi:10.1016/j.aop.2005.03.008 (2005) (cit. on p. 10).

[53] Elitzur, S.: *Impossibility of spontaneously breaking local symmetries*. Phys. Rev. D **12**, 3978–3982. doi:10.1103/PhysRevD.12.3978 (1975) (cit. on p. 10).

[54] Volovik, G. E.: *The universe in a helium droplet* 1. publ. ISBN: 0-19-850782-8 (Clarendon Press, Oxford [u.a.], 2003) (cit. on p. 11).

[55] Volovik, G. E.: *Topology of quantum vacuum* arXiv:1111.4627 [cond-mat, physics:gr-qc, physics:hep-ph]. (2012). doi:10.48550/arXiv.1111.4627 (cit. on p. 11).

[56] Tsuei, C. C. & Kirtley, J. R.: *Pairing symmetry in cuprate superconductors*. Rev. Mod. Phys. **72**, 969–1016. doi:10.1103/RevModPhys.72.969 (2000) (cit. on pp. 11, 24, 26, 55).

[57] Anderson, P. W.: *Plasmons, Gauge Invariance, and Mass*. Phys. Rev. **130**, 439–442. doi:10.1103/PhysRev.130.439 (1963) (cit. on p. 11).

[58] Englert, F. & Brout, R.: *Broken Symmetry and the Mass of Gauge Vector Mesons*. Phys. Rev. Lett. **13**, 321–323. doi:10.1103/PhysRevLett.13.321 (1964) (cit. on p. 11).

[59] Guralnik, G. S., Hagen, C. R. & Kibble, T. W. B.: *Global Conservation Laws and Massless Particles*. Phys. Rev. Lett. **13**, 585–587. doi:10.1103/PhysRevLett.13.585 (1964) (cit. on p. 11).

[60] Higgs, P. W.: *Broken symmetries, massless particles and gauge fields*. Physics Letters **12**, 132–133. doi:10.1016/0031-9163(64)91136-9 (1964) (cit. on p. 11).

[61] Higgs, P. W.: *Broken Symmetries and the Masses of Gauge Bosons*. Phys. Rev. Lett. **13**, 508–509. doi:10.1103/PhysRevLett.13.508 (1964) (cit. on p. 11).

[62] Timm, C.: *Theory of Superconductivity (Lecture Notes Wintersemester 2017/2018)* URL: https://tu-dresden.de/mn/physik/itp/cmt/ressourcen/dateien/skripte/Skript_Supra.pdf?lang=en (visited on 28.9.2022) (cit. on pp. 11, 13–15).

[63] Littlewood, P. B. & Varma, C. M.: *Amplitude collective modes in superconductors and their coupling to charge-density waves*. Phys. Rev. B **26**, 4883–4893. doi:10.1103/PhysRevB.26.4883 (1982) (cit. on p. 12).

[64] Sherman, D., Pracht, U. S., Gorshunov, B., Poran, S., Jesudasan, J., Chand, M., Raychaudhuri, P., Swanson, M., Trivedi, N., Auerbach, A., Scheffler, M., Frydman, A. & Dressel, M.: *The Higgs mode in disordered superconductors close to a quantum phase transition*. Nature Physics **11**, 188–192. doi:10.1038/nphys3227 (2015) (cit. on pp. 12, 83).

[65] Migdal, A.: *Interaction between electrons and lattice vibrations in a normal metal*. Sov. Phys. JETP **7**, 996–1001 (1958) (cit. on pp. 12, 17).

[66] Eliashberg, G.: *Temperature Green's function for electrons in a superconductor*. Sov. Phys. JETP **12**, 1000–1002 (1961) (cit. on p. 12).

[67] Bogoljubov, N. N.: *On a new method in the theory of superconductivity*. Nuovo Cim **7**, 794–805. doi:10.1007/BF02745585 (1958) (cit. on p. 13).

[68] Valatin, J. G.: *Comments on the theory of superconductivity*. Nuovo Cim **7**, 843–857. doi:10.1007/BF02745589 (1958) (cit. on p. 13).

[69] Gor'kov, L. P.: *Microscopic derivation of the Ginzburg-Landau equations in the theory of superconductivity*. Sov. Phys. JETP **9**, 1364–1367 (1959) (cit. on p. 14).

[70] Hlobil, P.: *Inelastic tunneling in superconducting junctions* doi:10.5445/IR/1000055763 (KIT-Bibliothek, Karlsruhe, 2016) (cit. on pp. 15, 18, 30, 33, 35, 36).

[71] Giaever, I., Hart, H. R. & Megerle, K.: *Tunneling into Superconductors at Temperatures below 1°K*. Phys. Rev. **126**, 941–948. doi:10.1103/PhysRev.126.941 (1962) (cit. on p. 18).

[72] McMillan, W. L. & Rowell, J. M.: *Lead Phonon Spectrum Calculated from Superconducting Density of States*. Phys. Rev. Lett. **14**, 108–112. doi:10.1103/PhysRevLett.14.108 (1965) (cit. on pp. 18, 37).

[73] Schackert, M., Märkl, T., Jandke, J., Hölzer, M., Ostanin, S., Gross, E. K. U., Ernst, A. & Wulfhekel, W.: *Local Measurement of the Eliashberg Function of Pb Islands: Enhancement of Electron-Phonon Coupling by Quantum Well States*. Phys. Rev. Lett. **114**, 047002. doi:10.1103/PhysRevLett.114.047002 (2015) (cit. on pp. 18, 33, 35).

[74] Jandke, J., Hlobil, P., Schackert, M., Wulfhekel, W. & Schmalian, J.: *Coupling to real and virtual phonons in tunneling spectroscopy of superconductors*. Phys. Rev. B **93**, 060505. doi:10.1103/PhysRevB.93.060505 (2016) (cit. on pp. 18, 35).

[75] Terashima, K., Sekiba, Y., Bowen, J. H., Nakayama, K., Kawahara, T., Sato, T., Richard, P., Xu, Y.-M., Li, L. J., Cao, G. H., Xu, Z.-A., Ding, H. & Takahashi, T.: *Fermi surface nesting induced strong pairing in iron-based superconductors*. Proceedings of the National Academy of Sciences **106**, 7330–7333. doi:10.1073/pnas.0900469106 (2009) (cit. on p. 19).

[76] Yin, J.-X., Wu, X.-X., Li, J., Wu, Z., Wang, J.-H., Ting, C.-S., Hor, P.-H., Liang, X. J., Zhang, C. L., Dai, P. C., Wang, X. C., Jin, C. Q., Chen, G. F., Hu, J. P., Wang, Z.-Q., Li, A., Ding, H. & Pan, S. H.: *Orbital selectivity of layer-resolved tunneling in the iron-based superconductor* $Ba_{0.6}K_{0.4}Fe_2As_2$. Phys. Rev. B **102**, 054515. doi:10.1103/PhysRevB.102.054515 (2020) (cit. on p. 19).

[77] Böhmer, A. E. & Kreisel, A.: *Nematicity, magnetism and superconductivity in FeSe*. J. Phys.: Condens. Matter **30**, 023001. doi:10.1088/1361-648X/aa9caa (2017) (cit. on p. 19).

[78] Shen, L. Y. L., Senozan, N. M. & Phillips, N. E.: *Evidence for Two Energy Gaps in High-Purity Superconducting Nb, Ta, and V*. Phys. Rev. Lett. **14**, 1025–1027. doi:10.1103/PhysRevLett.14.1025 (1965) (cit. on p. 19).

[79] Anderson, A. C., Satterthwaite, C. B. & Smith, S. C.: *Thermal Conductivity of Superconducting Niobium*. Phys. Rev. B **3**, 3762–3764. doi:10.1103/PhysRevB.3.3762 (1971) (cit. on p. 19).

[80] Almond, D. P., Lea, M. J. & Dobbs, E. R.: *Ultrasonic Evidence Against Multiple Energy Gaps in Superconducting Niobium*. Phys. Rev. Lett. **29**, 764–767. doi:10.1103/PhysRevLett.29.764 (1972) (cit. on p. 19).

[81] Sellers, G. J., Anderson, A. C. & Birnbaum, H. K.: *Anomalous heat capacities of niobium and tantalum below 1 K*. Phys. Rev. B **10**, 2771–2776. doi:10.1103/PhysRevB.10.2771 (1974) (cit. on p. 19).

[82] Floris, A., Sanna, A., Massidda, S. & Gross, E. K. U.: *Two-band superconductivity in Pb from ab initio calculations*. Phys. Rev. B **75**, 054508. doi:10.1103/PhysRevB.75.054508 (2007) (cit. on p. 19).

[83] Triola, C. & Balatsky, A. V.: *Pair symmetry conversion in driven multiband superconductors*. Phys. Rev. B **95**, 224518. doi:10.1103/PhysRevB.95.224518 (2017) (cit. on p. 19).

[84] Golubov, A. A. & Mazin, I. I.: *Effect of magnetic and nonmagnetic impurities on highly anisotropic superconductivity*. Phys. Rev. B **55**, 15146–15152. doi:10.1103/PhysRevB.55.15146 (1997) (cit. on p. 19).

[85] Komendová, L., Balatsky, A. V. & Black-Schaffer, A. M.: *Experimentally observable signatures of odd-frequency pairing in multiband superconductors*. Phys. Rev. B **92**, 094517. doi:10.1103/PhysRevB.92.094517 (2015) (cit. on p. 20).

[86] Tanaka, Y., Sato, M. & Nagaosa, N.: *Symmetry and Topology in Superconductors –Odd-Frequency Pairing and Edge States–*. J. Phys. Soc. Jpn. **81**, 011013. doi:10.1143/JPSJ.81.011013 (2012) (cit. on p. 20).

[87] Ichioka, M., Kogan, V. G. & Schmalian, J.: *Locking of length scales in two-band superconductors*. Phys. Rev. B **95**, 064512. doi:10.1103/PhysRevB.95.064512 (2017) (cit. on pp. 20, 105).

[88] Deutscher, G.: *New Superconductors : From Granular to High T_c* ISBN: 9810230893 (World Scientific, Singapore, 2006) (cit. on pp. 20, 22).

[89] Averin, D. V. & Likharev, K. K.: *Coulomb blockade of single-electron tunneling, and coherent oscillations in small tunnel junctions*. J Low Temp Phys **62**, 345–373. doi:10.1007/BF00683469 (1986) (cit. on p. 22).

[90] Gross, R. & Marx, A.: *Applied Superconductivity: Josephson Effect and Superconducting Electronics (Lecture Notes October, 2005)* URL: https://www.wmi.badw.de/fileadmin/WMI/Lecturenotes/Applied_Superconductivity/AS_Chapter3.pdf (visited on 23.8.2022) (cit. on p. 22).

[91] Goll, G.: *Unconventional Superconductors : Experimental Investigation of the Order-Parameter Symmetry* ISBN: 978-3-540-31500-1 (Springer Berlin Heidelberg, 2006) (cit. on p. 23).

[92] Scalapino, D. J.: *The case for $d_{x^2-y^2}$ pairing in the cuprate superconductors*. Physics Reports **250**, 329–365. doi:10.1016/0370-1573(94)00086-I (1995) (cit. on pp. 23, 55).

[93] Sigrist, M. & Ueda, K.: *Phenomenological theory of unconventional superconductivity*. Rev. Mod. Phys. **63**, 239–311. doi:10.1103/RevModPhys.63.239 (1991) (cit. on pp. 24, 25).

[94] Annett, J. F.: *Symmetry of the order parameter for high-temperature superconductivity*. Advances in Physics **39**, 83–126. doi:10.1080/00018739000101481 (1990) (cit. on p. 25).

[95] Wenger, F. & Östlund, S.: *d -wave pairing in tetragonal superconductors*. Phys. Rev. B **47**, 5977–5983. doi:10.1103/PhysRevB.47.5977 (1993) (cit. on p. 26).

[96] Melnikov, N. B. & Reser, B. I.: *Dynamic Spin-Fluctuation Theory of Metallic Magnetism* ISBN: 978-3-319-92972-9 978-3-319-92974-3. doi:10.1007/978-3-319-92974-3 (Springer International Publishing, Cham, 2018) (cit. on p. 27).

[97] Chubukov, A. V.: *Theory of the leading edge gap in underdoped cuprates*. EPL **44**, 655. doi:10.1209/epl/i1998-00522-9 (1998) (cit. on p. 28).

[98] Chubukov, A. V. & Schmalian, J.: *Superconductivity due to massless boson exchange in the strong-coupling limit*. Phys. Rev. B **72**, 174520. doi:10.1103/PhysRevB.72.174520 (2005) (cit. on pp. 29, 83, 87).

[99] Bardeen, J.: *Tunnelling from a Many-Particle Point of View*. Phys. Rev. Lett. **6**, 57–59. doi:10.1103/PhysRevLett.6.57 (1961) (cit. on p. 32).

[100] Tersoff, J. & Hamann, D. R.: *Theory of the scanning tunneling microscope*. Phys. Rev. B **31**, 805–813. doi:10.1103/PhysRevB.31.805 (1985) (cit. on p. 32).

[101] Berthod, C. & Giamarchi, T.: *Tunneling conductance and local density of states in tight-binding junctions*. Phys. Rev. B **84**, 155414. doi:10.1103/PhysRevB.84.155414 (2011) (cit. on p. 33).

[102] Cohen, M. H., Falicov, L. M. & Phillips, J. C.: *Superconductive Tunneling*. Phys. Rev. Lett. **8**, 316–318. doi:10.1103/PhysRevLett.8.316 (1962) (cit. on p. 33).

[103] Young, R., Ward, J. & Scire, F.: *The Topografiner: An Instrument for Measuring Surface Microtopography*. Review of Scientific Instruments **43**, 999–1011. doi:10.1063/1.1685846 (1972) (cit. on p. 39).

[104] Binnig, G., Rohrer, H., Gerber, C. & Weibel, E.: *Surface Studies by Scanning Tunneling Microscopy*. Phys. Rev. Lett. **49**, 57–61. doi:10.1103/PhysRevLett.49.57 (1982) (cit. on p. 40).

[105] Renner, C., Niedermann, P., Kent, A. D. & Fischer, O.: *A vertical piezoelectric inertial slider*. Review of Scientific Instruments **61**, 965–967. doi:10.1063/1.1141450 (1990) (cit. on p. 40).

[106] Jäck, B., Eltschka, M., Assig, M., Etzkorn, M., Ast, C. R. & Kern, K.: *Critical Josephson current in the dynamical Coulomb blockade regime*. Phys. Rev. B **93**, 020504. doi:10.1103/PhysRevB.93.020504 (2016) (cit. on pp. 42, 76).

[107] Zhang, L., Miyamachi, T., Tomanić, T., Dehm, R. & Wulfhekel, W.: *A compact sub-Kelvin ultrahigh vacuum scanning tunneling microscope with high energy resolution and high stability*. Rev. Sci. Instrum. **82**, 103702. doi:10.1063/1.3646468 (2011) (cit. on p. 43).

[108] Balashov, T., Meyer, M. & Wulfhekel, W.: *A compact ultrahigh vacuum scanning tunneling microscope with dilution refrigeration*. Review of Scientific Instruments **89**, 113707. doi:10.1063/1.5043636 (2018) (cit. on pp. 43, 45).

[109] Pobell, F.: *Matter and methods at low temperatures : with 28 tables, and 81 problems* 3., rev. and expanded ed., 149–189. ISBN: 9783540463566 (Springer, Berlin, 2007) (cit. on p. 46).

[110] Ibe, J. P., Bey, P. P., Brandow, S. L., Brizzolara, R. A., Burnham, N. A., DiLella, D. P., Lee, K. P., Marrian, C. R. K. & Colton, R. J.: *On the electrochemical etching of tips for scanning tunneling microscopy*. Journal of Vacuum Science & Technology A **8**, 3570–3575. doi:10.1116/1.576509 (1990) (cit. on p. 47).

[111] Méndez, J., Luna, M. & Baró, A. M.: *Preparation of STM W tips and characterization by FEM, TEM and SEM*. Surface Science **266**, 294–298. doi:10.1016/0039-6028(92)91036-B (1992) (cit. on p. 48).

[112] Ekvall, I., Wahlström, E., Claesson, D., Olin, H. & Olsson, E.: *Preparation and characterization of electrochemically etched W tips for STM*. Meas. Sci. Technol. **10**, 11–18. doi:10.1088/0957-0233/10/1/006 (1999) (cit. on pp. 48, 74).

[113] Hammer, B. & Norskov, J. K.: *Why gold is the noblest of all the metals*. Nature **376**, 238–240. doi:10.1038/376238a0 (1995) (cit. on p. 48).

[114] Wu, M. K., Ashburn, J. R., Torng, C. J., Hor, P. H., Meng, R. L., Gao, L., Huang, Z. J., Wang, Y. Q. & Chu, C. W.: *Superconductivity at 93 K in a new mixed-phase Y-Ba-Cu-O compound system at ambient pressure*. Phys. Rev. Lett. **58**, 908–910. doi:10.1103/PhysRevLett.58.908 (1987) (cit. on p. 51).

[115] Maeda, H., Tanaka, Y., Fukutomi, M. & Asano, T.: *A New High-Tc Oxide Superconductor without a Rare Earth Element*. Jpn. J. Appl. Phys. **27**, L209. doi:10.1143/JJAP.27.L209 (1988) (cit. on p. 51).

[116] Hervieu, M., Michel, C., Domenges, B., Laligant, Y., Lebail, A., Ferey, G. & Raveau, B.: *ELECTRON MICROSCOPY STUDY OF THE SUPERCONDUCTOR "$Bi_2Sr_2CaCu_2O_{8+\delta}$"*. Mod. Phys. Lett. B **02**, 491–500. doi:10.1142/S0217984988000060 (1988) (cit. on p. 52).

[117] Williams, A., Kwei, G. H., Von Dreele, R. B., Raistrick, I. D. & Bish, D. L.: *Joint x-ray and neutron refinement of the structure of superconducting $YBa_2Cu_3O_{7-x}$: Precision structure, anisotropic thermal parameters, strain, and cation disorder.* Phys. Rev. B **37**, 7960–7962. doi:10.1103/PhysRevB.37.7960 (1988) (cit. on p. 52).

[118] Pickett, W. E.: *Electronic structure of the high-temperature oxide superconductors.* Rev. Mod. Phys. **61**, 433–512. doi:10.1103/RevModPhys.61.433 (1989) (cit. on pp. 51, 52).

[119] Pan, S. H., Hudson, E. W., Ma, J. & Davis, J. C.: *Imaging and identification of atomic planes of cleaved $Bi_2Sr_2CaCu_2O_{8+\delta}$ by high resolution scanning tunneling microscopy.* Appl. Phys. Lett. **73**, 58–60. doi:10.1063/1.121722 (1998) (cit. on p. 52).

[120] Fischer, Ø., Kugler, M., Maggio-Aprile, I., Berthod, C. & Renner, C.: *Scanning tunneling spectroscopy of high-temperature superconductors.* Rev. Mod. Phys. **79**, 353–419. doi:10.1103/RevModPhys.79.353 (2007) (cit. on pp. 52, 58, 60, 67).

[121] Rossat-Mignod, J., Regnault, L. P., Vettier, C., Bourges, P., Burlet, P., Bossy, J., Henry, J. Y. & Lapertot, G.: *Neutron scattering study of the $YBa_2Cu_3O_{6+x}$ system.* Physica C: Superconductivity **185-189**, 86–92. doi:10.1016/0921-4534(91)91955-4 (1991) (cit. on pp. 54, 66).

[122] Presland, M., Tallon, J., Buckley, R., Liu, R. & Flower, N.: *General trends in oxygen stoichiometry effects on Tc in Bi and Tl superconductors.* Physica C: Superconductivity **176**, 95–105. doi:10.1016/0921-4534(91)90700-9 (1991) (cit. on p. 53).

[123] Obertelli, S. D., Cooper, J. R. & Tallon, J. L.: *Systematics in the thermoelectric power of high- T c oxides.* Phys. Rev. B **46**, 14928–14931. doi:10.1103/PhysRevB.46.14928 (1992) (cit. on p. 53).

[124] Honma, T. & Hor, P. H.: *Unified electronic phase diagram for hole-doped high-T_c cuprates.* Phys. Rev. B **77**, 184520. doi:10.1103/PhysRevB.77.184520 (2008) (cit. on p. 53).

[125] Drozdov, I. K., Pletikosić, I., Kim, C.-K., Fujita, K., Gu, G. D., Davis, J. C. S., Johnson, P. D., Božović, I. & Valla, T.: *Phase diagram of $Bi_2Sr_2CaCu_2O_{8+\delta}$ revisited.* Nat. Commun. **9**, 5210. doi:10.1038/s41467-018-07686-w (2018) (cit. on p. 53).

[126] Sato, Y., Kasahara, S., Murayama, H., Kasahara, Y., Moon, E.-G., Nishizaki, T., Loew, T., Porras, J., Keimer, B., Shibauchi, T. & Matsuda, Y.: *Thermodynamic evidence for a nematic phase transition at the onset of the pseudogap in $YBa_2Cu_3O_y$.* Nature Phys **13**, 1074–1078. doi:10.1038/nphys4205 (2017) (cit. on pp. 53, 54).

[127] Yang, K.-Y., Rice, T. M. & Zhang, F.-C.: *A Phenomenological Theory of The Pseudogap State*. Phys. Rev. B **73**, 174501. doi:10.1103/PhysRevB.73.174501 (2006) (cit. on p. 54).

[128] Barrett, S. E., Durand, D. J., Pennington, C. H., Slichter, C. P., Friedmann, T. A., Rice, J. P. & Ginsberg, D. M.: *^{63}Cu Knight shifts in the superconducting state of $YBa_2Cu_3O_{7-\delta}$ ($T_c = 90K$)*. Phys. Rev. B **41**, 6283–6296. doi:10.1103/PhysRevB.41.6283 (1990) (cit. on p. 55).

[129] Damascelli, A., Hussain, Z. & Shen, Z.-X.: *Angle-resolved photoemission studies of the cuprate superconductors*. Rev. Mod. Phys. **75**, 473–541. doi:10.1103/RevModPhys.75.473 (2003) (cit. on pp. 55, 61).

[130] Yu, G., Li, Y., Motoyama, E. M. & Greven, M.: *A universal relationship between magnetic resonance and superconducting gap in unconventional superconductors*. Nature Phys **5**, 873–875. doi:10.1038/nphys1426 (2009) (cit. on pp. 56, 58, 66, 67).

[131] Sui, X.-H., Tang, H., Zhao, S. P. & Su, Z.-B.: *Eliashberg analysis of $Bi_2Sr_2CaCu_2O_{8+\delta}$ intrinsic tunneling spectra*. Physica C: Superconductivity and its Applications **511**, 15–19. doi:10.1016/j.physc.2015.02.005 (2015) (cit. on p. 58).

[132] Chubukov, A. V. & Schmalian, J.: *Pairing glue in cuprate superconductors from the self-energy revealed via machine learning*. Phys. Rev. B **101**, 180510. doi:10.1103/PhysRevB.101.180510 (2020) (cit. on p. 58).

[133] Yamaji, Y., Yoshida, T., Fujimori, A. & Imada, M.: *Hidden self-energies as origin of cuprate superconductivity revealed by machine learning*. Phys. Rev. Research **3**, 043099. doi:10.1103/PhysRevResearch.3.043099 (2021) (cit. on p. 58).

[134] Maggio-Aprile, I., Renner, C., Erb, A., Walker, E. & Fischer, Ø.: *Direct Vortex Lattice Imaging and Tunneling Spectroscopy of Flux Lines on $YBa_2Cu_3O_{7-\delta}$*. Phys. Rev. Lett. **75**, 2754–2757. doi:10.1103/PhysRevLett.75.2754 (1995) (cit. on p. 61).

[135] Miller, T. G., McElfresh, M. & Reifenberger, R.: *Density-of-states fine structure in the tunneling conductance of Y-Ba-Cu-O films: A comparison between experiment and theory*. Phys. Rev. B **48**, 7499–7504. doi:10.1103/PhysRevB.48.7499 (1993) (cit. on p. 61).

[136] Tachiki, M., Takahashi, S., Steglich, F. & Adrian, H.: *Tunneling conductance in layered oxide superconductors*. Z. Physik B - Condensed Matter **80**, 161–166. doi:10.1007/BF01357497 (1990) (cit. on p. 61).

[137] Dynes, R. C., Narayanamurti, V. & Garno, J. P.: *Direct Measurement of Quasiparticle-Lifetime Broadening in a Strong-Coupled Superconductor*. Phys. Rev. Lett. **41**, 1509–1512. doi:10.1103/PhysRevLett.41.1509 (1978) (cit. on p. 61).

[138] Norman, M. R., Randeria, M., Ding, H. & Campuzano, J. C.: *Phenomenological Models for the Gap Anisotropy of Bi-2212 as Measured by ARPES*. Phys. Rev. B **52**, 615–622. doi:10.1103/PhysRevB.52.615 (1995) (cit. on p. 62).

[139] Sassa, Y., Radović, M., Månsson, M., Razzoli, E., Cui, X. Y., Pailhès, S., Guerrero, S., Shi, M., Willmott, P. R., Miletto Granozio, F., Mesot, J., Norman, M. R. & Patthey, L.: *Ortho-II band folding in $YBa_2Cu_3O_{7-\delta}$ films revealed by angle-resolved photoemission*. Phys. Rev. B **83**, 140511. doi:10.1103/PhysRevB.83.140511 (2011) (cit. on p. 62).

[140] Gold, R.: *An iterative unfolding method for response matrices* tech. rep. ANL-6984 (Argonne National Laboratories, Argonne, IL, 1964) (cit. on p. 65).

[141] Morháč, M., Kliman, J., Matoušek, V., Veselský, M. & Turzo, I.: *Efficient one- and two-dimensional gold deconvolution and its application to γ-ray spectra decomposition*. Nuclear Instruments and Methods in Physics Research Section A: Accelerators, Spectrometers, Detectors and Associated Equipment **401**, 385–408. doi:10.1016/S0168-9002(97)01058-9 (1997) (cit. on pp. 65, 175).

[142] Hwang, J., Timusk, T., Schachinger, E. & Carbotte, J. P.: *Evolution of the bosonic spectral density of the high-temperature superconductor $Bi_2Sr_2CaCu_2O_{8+\delta}$*. Phys. Rev. B **75**, 144508. doi:10.1103/PhysRevB.75.144508 (2007) (cit. on p. 66).

[143] Pailhès, S., Ulrich, C., Fauqué, B., Hinkov, V., Sidis, Y., Ivanov, A., Lin, C. T., Keimer, B. & Bourges, P.: *Doping Dependence of Bilayer Resonant Spin Excitations in $(Y,Ca)Ba_2Cu_3O_{6+x}$*. Phys. Rev. Lett. **96**, 257001. doi:10.1103/PhysRevLett.96.257001 (2006) (cit. on p. 66).

[144] Pailhès, S., Sidis, Y., Bourges, P., Hinkov, V., Ivanov, A., Ulrich, C., Regnault, L. P. & Keimer, B.: *Resonant Magnetic Excitations at High Energy in Superconducting $YBa_2Cu_3O_{6.85}$*. Phys. Rev. Lett. **93**, 167001. doi:10.1103/PhysRevLett.93.167001 (2004) (cit. on p. 66).

[145] Pailhès, S., Sidis, Y., Bourges, P., Ulrich, C., Hinkov, V., Regnault, L. P., Ivanov, A., Liang, B., Lin, C. T., Bernhard, C. & Keimer, B.: *Two Resonant Magnetic Modes in an Overdoped High T_c Superconductor*. Phys. Rev. Lett. **91**, 237002. doi:10.1103/PhysRevLett.91.237002 (2003) (cit. on p. 66).

[146] Yang, F., Gozlinski, T., Storbeck, T., Grünhaupt, L., Pop, I. M. & Wulfhekel, W.: *Microscopic charging and in-gap states in superconducting granular aluminum*. Phys. Rev. B **102**, 104502. doi:10.1103/PhysRevB.102.104502 (2020) (cit. on pp. 69, 73–79, 81, 86, 133).

[147] Abeles, B., Cohen, R. W. & Stowell, W. R.: *Critical Magnetic Fields of Granular Superconductors*. Phys. Rev. Lett. **18**, 902–905. doi:10.1103/PhysRevLett.18.902 (1967) (cit. on p. 70).

[148] Cohen, R. W. & Abeles, B.: *Superconductivity in Granular Aluminum Films.* Phys. Rev. **168**, 444–450. doi:10.1103/PhysRev.168.444 (1968) (cit. on pp. 70, 73).

[149] Pracht, U. S., Bachar, N., Benfatto, L., Deutscher, G., Farber, E., Dressel, M. & Scheffler, M.: *Enhanced Cooper pairing versus suppressed phase coherence shaping the superconducting dome in coupled aluminum nanograins.* Phys. Rev. B **93**, 100503. doi:10.1103/PhysRevB.93.100503 (2016) (cit. on pp. 70, 71, 83, 86, 90).

[150] Deutscher, G., Fenichel, H., Gershenson, M., Grünbaum, E. & Ovadyahu, Z.: *Transition to zero dimensionality in granular aluminum superconducting films.* J Low Temp Phys **10**, 231–243. doi:10.1007/BF00655256 (1973) (cit. on pp. 70, 73, 74, 77).

[151] Dynes, R. C. & Garno, J. P.: *Metal-Insulator Transition in Granular Aluminum.* Phys. Rev. Lett. **46**, 137–140. doi:10.1103/PhysRevLett.46.137 (1981) (cit. on p. 70).

[152] Dynes, R. C., Garno, J. P., Hertel, G. B. & Orlando, T. P.: *Tunneling Study of Superconductivity near the Metal-Insulator Transition.* Phys. Rev. Lett. **53**, 2437–2440. doi:10.1103/PhysRevLett.53.2437 (1984) (cit. on p. 70).

[153] Levy-Bertrand, F., Klein, T., Grenet, T., Dupré, O., Benoît, A., Bideaud, A., Bourrion, O., Calvo, M., Catalano, A., Gomez, A., Goupy, J., Grünhaupt, L., Luepke, U. v., Maleeva, N., Valenti, F., Pop, I. M. & Monfardini, A.: *Electrodynamics of granular aluminum from superconductor to insulator: Observation of collective superconducting modes.* Phys. Rev. B **99**, 094506. doi:10.1103/PhysRevB.99.094506 (2019) (cit. on pp. 70, 71, 74, 75, 79–81, 83).

[154] Parmenter, R. H.: *Size Effect in a Granular Superconductor.* Phys. Rev. **166**, 392–396. doi:10.1103/PhysRev.166.392 (1968) (cit. on p. 71).

[155] Kresin, V. Z. & Ovchinnikov, Y. N.: *Shell structure and strengthening of superconducting pair correlation in nanoclusters.* Phys. Rev. B **74**, 024514. doi:10.1103/PhysRevB.74.024514 (2006) (cit. on p. 71).

[156] García-García, A. M., Urbina, J. D., Yuzbashyan, E. A., Richter, K. & Altshuler, B. L.: *BCS superconductivity in metallic nanograins: Finite-size corrections, low-energy excitations, and robustness of shell effects.* Phys. Rev. B **83**, 014510. doi:10.1103/PhysRevB.83.014510 (2011) (cit. on p. 71).

[157] Klein, J. & Leger, A.: *Experimental evidence for superconductivity enhancement mechanism.* Physica **55**, 740–743. doi:10.1016/0031-8914(71)90327-2 (1971) (cit. on p. 71).

[158] Houben, K., Jochum, J. K., Couet, S., Menéndez, E., Picot, T., Hu, M. Y., Zhao, J. Y., Alp, E. E., Vantomme, A., Temst, K. & Van Bael, M. J.: *The influence of phonon softening on the superconducting critical temperature of Sn nanostructures*. Sci Rep **10**, 5729. doi:10.1038/s41598-020-62617-4 (2020) (cit. on p. 71).

[159] Mayoh, J. & García-García, A. M.: *Strong enhancement of bulk superconductivity by engineered nanogranularity*. Phys. Rev. B **90**, 134513. doi:10.1103/PhysRevB.90.134513 (2014) (cit. on p. 71).

[160] Abeles, B.: *Effect of charging energy on superconductivity in granular metal films*. Phys. Rev. B **15**, 2828–2829. doi:10.1103/PhysRevB.15.2828 (1977) (cit. on p. 71).

[161] Deutscher, G.: *Critical fluctuations in granular superconductors*. Physics Letters A **35**, 28–29. doi:10.1016/0375-9601(71)90017-X (1971) (cit. on pp. 71, 87).

[162] Deutscher, G., Imry, Y. & Gunther, L.: *Superconducting phase transitions in granular systems*. Phys. Rev. B **10**, 4598–4606. doi:10.1103/PhysRevB.10.4598 (1974) (cit. on p. 71).

[163] Moshe, A., Bachar, N., Lerer, S., Lereah, Y. & Deutscher, G.: *Multi-Level Kondo Effect and Enhanced Critical Temperature in Nanoscale Granular Al*. J Supercond Nov Magn **31**, 733–736. doi:10.1007/s10948-017-4330-z (2018) (cit. on pp. 72, 79).

[164] Choi, M.-S., Lee, M., Kang, K. & Belzig, W.: *Kondo effect and Josephson current through a quantum dot between two superconductors*. Phys. Rev. B **70**, 020502. doi:10.1103/PhysRevB.70.020502 (2004) (cit. on p. 72).

[165] Deutscher, G.: *Granular Superconductivity: a Playground for Josephson, Anderson, Kondo, and Mott*. J Supercond Nov Magn **34**, 1699–1703. doi:10.1007/s10948-020-05773-y (2021) (cit. on p. 72).

[166] Moshe, A. G., Farber, E. & Deutscher, G.: *Optical conductivity of granular aluminum films near the Mott metal-to-insulator transition*. Phys. Rev. B **99**, 224503. doi:10.1103/PhysRevB.99.224503 (2019) (cit. on p. 72).

[167] Haussmann, R.: *Crossover from BCS superconductivity to Bose-Einstein condensation: A self-consistent theory*. Z. Physik B - Condensed Matter **91**, 291–308. doi:10.1007/BF01344058 (1993) (cit. on p. 72).

[168] Pistolesi, F. & Strinati, G. C.: *Evolution from BCS superconductivity to Bose condensation: Role of the parameter $k_F\xi$*. Phys. Rev. B **49**, 6356–6359. doi:10.1103/PhysRevB.49.6356 (1994) (cit. on p. 72).

[169] Palestini, F. & Strinati, G. C.: *Temperature dependence of the pair coherence and healing lengths for a fermionic superfluid throughout the BCS-BEC crossover*. Phys. Rev. B **89**, 224508. doi:10.1103/PhysRevB.89.224508 (2014) (cit. on p. 72).

[170] Douglass, D. H. & Meservey, R.: *Energy Gap Measurements by Tunneling Between Superconducting Films. I. Temperature Dependence*. Phys. Rev. **135**, A19–A23. doi:10.1103/PhysRev.135.A19 (1964) (cit. on p. 73).

[171] Szkutnik, P. D., Piednoir, A., Ronda, A., Marchi, F., Tonneau, D., Dallaporta, H. & Hanbücken, M.: *STM studies: spatial resolution limits to fit observations in nanotechnology*. Applied Surface Science. *Surface Science in Micro & Nanotechnology* **164**, 169–174. doi:10.1016/S0169-4332(00)00337-8 (2000) (cit. on p. 74).

[172] Guise, O. L., Ahner, J. W., Jung, M.-C., Goughnour, P. C. & Yates, J. T.: *Reproducible Electrochemical Etching of Tungsten Probe Tips*. Nano Lett. **2**, 191–193. doi:10.1021/nl010094q (2002) (cit. on p. 74).

[173] Valenti, F., Henriques, F., Catelani, G., Maleeva, N., Grünhaupt, L., von Lüpke, U., Skacel, S. T., Winkel, P., Bilmes, A., Ustinov, A. V., Goupy, J., Calvo, M., Benoît, A., Levy-Bertrand, F., Monfardini, A. & Pop, I. M.: *Interplay Between Kinetic Inductance, Nonlinearity, and Quasiparticle Dynamics in Granular Aluminum Microwave Kinetic Inductance Detectors*. Phys. Rev. Applied **11**, 054087. doi:10.1103/PhysRevApplied.11.054087 (2019) (cit. on pp. 74, 75).

[174] Mott, N. F.: *Metal-Insulator Transition*. Rev. Mod. Phys. **40**, 677–683. doi:10.1103/RevModPhys.40.677 (1968) (cit. on p. 76).

[175] Ast, C. R., Jäck, B., Senkpiel, J., Eltschka, M., Etzkorn, M., Ankerhold, J. & Kern, K.: *Sensing the quantum limit in scanning tunnelling spectroscopy*. Nat Commun **7**, 13009. doi:10.1038/ncomms13009 (2016) (cit. on p. 76).

[176] Feinberg, D.: *Andreev scattering and cotunneling between two superconductor-normal metal interfaces: the dirty limit*. The European Physical Journal B - Condensed Matter **36**, 419–422. doi:10.1140/epjb/e2003-00361-6 (2003) (cit. on p. 76).

[177] Falci, G., Feinberg, D. & Hekking, F. W. J.: *Correlated tunneling into a superconductor in a multiprobe hybrid structure*. Europhys. Lett. **54**, 255–261. doi:10.1209/epl/i2001-00303-0 (2001) (cit. on p. 76).

[178] Russo, S., Kroug, M., Klapwijk, T. M. & Morpurgo, A. F.: *Experimental Observation of Bias-Dependent Nonlocal Andreev Reflection*. Phys. Rev. Lett. **95**, 027002. doi:10.1103/PhysRevLett.95.027002 (2005) (cit. on p. 76).

[179] Lerer, S., Bachar, N., Deutscher, G. & Dagan, Y.: *Nernst effect beyond the coherence critical field of a nanoscale granular superconductor*. Phys. Rev. B **90**, 214521. doi:10.1103/PhysRevB.90.214521 (2014) (cit. on p. 77).

[180] Tinkham, M.: *Metallic quantum dots*. Philosophical Magazine B **79**, 1267–1280. doi:10.1080/13642819908216970 (1999) (cit. on p. 77).

[181] Hong, I.-P., Brun, C., Pivetta, M., Patthey, F. & Schneider, W.-D.: *Coulomb blockade phenomena observed in supported metallic nanoislands*. Frontiers in Physics **1**, 13 (2013) (cit. on p. 78).

[182] Bachar, N., Levy, A., Prokscha, T., Suter, A., Morenzoni, E., Salman, Z. & Deutscher, G.: *Kubo spins in nanoscale aluminum grains: A muon spin relaxation study*. Phys. Rev. B **101**, 024424. doi:10.1103/PhysRevB.101.024424 (2020) (cit. on p. 80).

[183] Farinacci, L., Ahmadi, G., Reecht, G., Ruby, M., Bogdanoff, N., Peters, O., Heinrich, B. W., von Oppen, F. & Franke, K. J.: *Tuning the Coupling of an Individual Magnetic Impurity to a Superconductor: Quantum Phase Transition and Transport*. Phys. Rev. Lett. **121**, 196803. doi:10.1103/PhysRevLett.121.196803 (2018) (cit. on p. 80).

[184] Pracht, U. S., Cea, T., Bachar, N., Deutscher, G., Farber, E., Dressel, M., Scheffler, M., Castellani, C., García-García, A. M. & Benfatto, L.: *Optical signatures of the superconducting Goldstone mode in granular aluminum: Experiments and theory*. Phys. Rev. B **96**, 094514. doi:10.1103/PhysRevB.96.094514 (2017) (cit. on pp. 83, 90).

[185] Klein, J. & Leger, A.: *Tunneling measurement of phonon spectrum in granular Al*. Physics Letters A **28**, 134–135. doi:10.1016/0375-9601(68)90431-3 (1968) (cit. on p. 83).

[186] Frederick, B. G., Lee, M. B. & Richardson, N. V.: *A vibrational characterisation of the O/A1(111) system: a reassignment of HREELS data*. Surface Science **348**, L71–L74. doi:10.1016/0039-6028(95)01101-3 (1996) (cit. on p. 83).

[187] Maleeva, N., Grünhaupt, L., Klein, T., Levy-Bertrand, F., Dupre, O., Calvo, M., Valenti, F., Winkel, P., Friedrich, F., Wernsdorfer, W., Ustinov, A. V., Rotzinger, H., Monfardini, A., Fistul, M. V. & Pop, I. M.: *Circuit quantum electrodynamics of granular aluminum resonators*. Nat Commun **9**. doi:10.1038/s41467-018-06386-9 (2018) (cit. on pp. 83, 88–90).

[188] Stern, F.: *Polarizability of a Two-Dimensional Electron Gas*. Phys. Rev. Lett. **18**, 546–548. doi:10.1103/PhysRevLett.18.546 (1967) (cit. on p. 83).

[189] Swanson, M., Loh, Y. L., Randeria, M. & Trivedi, N.: *Dynamical Conductivity across the Disorder-Tuned Superconductor-Insulator Transition*. Phys. Rev. X **4**, 021007. doi:10.1103/PhysRevX.4.021007 (2014) (cit. on p. 83).

[190] Nakamura, S., Iida, Y., Murotani, Y., Matsunaga, R., Terai, H. & Shimano, R.: *Infrared Activation of the Higgs Mode by Supercurrent Injection in Superconducting NbN*. Phys. Rev. Lett. **122**, 257001. doi:10.1103/PhysRevLett.122.257001 (2019) (cit. on p. 83).

[191] Cea, T., Castellani, C., Seibold, G. & Benfatto, L.: *Nonrelativistic Dynamics of the Amplitude (Higgs) Mode in Superconductors*. Phys. Rev. Lett. **115**, 157002. doi:10.1103/PhysRevLett.115.157002 (2015) (cit. on p. 83).

[192] Tien, P. K. & Gordon, J. P.: *Multiphoton Process Observed in the Interaction of Microwave Fields with the Tunneling between Superconductor Films*. Phys. Rev. **129**, 647–651. doi:10.1103/PhysRev.129.647 (1963) (cit. on pp. 87, 90).

[193] Cibert, C., Hidalgo, H., Champeaux, C., Tristant, P., Tixier, C., Desmaison, J. & Catherinot, A.: *Properties of aluminum oxide thin films deposited by pulsed laser deposition and plasma enhanced chemical vapor deposition*. Thin Solid Films **516**, 1290–1296. doi:10.1016/j.tsf.2007.05.064 (2008) (cit. on p. 89).

[194] Peters, O., Bogdanoff, N., Acero González, S., Melischek, L., Simon, J. R., Reecht, G., Winkelmann, C. B., von Oppen, F. & Franke, K. J.: *Resonant Andreev reflections probed by photon-assisted tunnelling at the atomic scale*. Nat. Phys. **16**, 1222–1226. doi:10.1038/s41567-020-0972-z (2020) (cit. on p. 90).

[195] Souquet, J.-R., Woolley, M. J., Gabelli, J., Simon, P. & Clerk, A. A.: *Photon-assisted tunnelling with nonclassical light*. Nat Commun **5**, 5562. doi:10.1038/ncomms6562 (2014) (cit. on p. 91).

[196] Krueger, W. H. & Pollack, S. R.: *The initial oxidation of aluminum thin films at room temperature*. Surface Science **30**, 263–279. doi:10.1016/0039-6028(72)90002-7 (1972) (cit. on p. 92).

[197] Cai, N., Zhou, G., Müller, K. & Starr, D. E.: *Effect of oxygen gas pressure on the kinetics of alumina film growth during the oxidation of Al(111) at room temperature*. Phys. Rev. B **84**, 125445. doi:10.1103/PhysRevB.84.125445 (2011) (cit. on p. 92).

[198] Bradshaw, A. M., Hofmann, P. & Wyrobisch, W.: *The interaction of oxygen with aluminium (111)*. Surface Science **68**, 269–276. doi:10.1016/0039-6028(77)90211-4 (1977) (cit. on p. 92).

[199] Cai, N., Zhou, G., Müller, K. & Starr, D. E.: *Tuning the Limiting Thickness of a Thin Oxide Layer on Al(111) with Oxygen Gas Pressure*. Phys. Rev. Lett. **107**, 035502. doi:10.1103/PhysRevLett.107.035502 (2011) (cit. on pp. 92, 94).

[200] Baran, J. D., Grönbeck, H. & Hellman, A.: *Mechanism for Limiting Thickness of Thin Oxide Films on Aluminum*. Phys. Rev. Lett. **112**, 146103. doi:10.1103/PhysRevLett.112.146103 (2014) (cit. on p. 94).

[201] Jeurgens, L. P. H., Sloof, W. G., Tichelaar, F. D. & Mittemeijer, E. J.: *Growth kinetics and mechanisms of aluminum-oxide films formed by thermal oxidation of aluminum*. Journal of Applied Physics **92**, 1649–1656. doi:10.1063/1.1491591 (2002) (cit. on p. 94).

[202] Deng, S., Viola, L. & Ortiz, G.: *Majorana Modes in Time-Reversal Invariant s -Wave Topological Superconductors*. Phys. Rev. Lett. **108**, 036803. doi:10.1103/PhysRevLett.108.036803 (2012) (cit. on p. 97).

[203] Deng, S., Ortiz, G. & Viola, L.: *Multiband s -wave topological superconductors: Role of dimensionality and magnetic field response*. Phys. Rev. B **87**, 205414. doi:10.1103/PhysRevB.87.205414 (2013) (cit. on p. 97).

[204] Vagov, A., Shanenko, A. A., Milošević, M. V., Axt, V. M., Vinokur, V. M., Aguiar, J. A. & Peeters, F. M.: *Superconductivity between standard types: Multiband versus single-band materials*. Phys. Rev. B **93**, 174503. doi:10.1103/PhysRevB.93.174503 (2016) (cit. on pp. 97, 113).

[205] Rex, S. & Willa, R.: *A topological flux trap: Majorana bound states at screw dislocations*. New J. Phys. **24**, 053057. doi:10.1088/1367-2630/ac61cf (2022) (cit. on p. 97).

[206] Heid, R., Bohnen, K.-P., Sklyadneva, I. Y. & Chulkov, E. V.: *Effect of spin-orbit coupling on the electron-phonon interaction of the superconductors Pb and Tl*. Phys. Rev. B **81**, 174527. doi:10.1103/PhysRevB.81.174527 (2010) (cit. on p. 97).

[207] Gozlinski, T., Li, Q., Heid, R., Nemoto, R., Willa, R., Yamada, T. K., Schmalian, J. & Wulfhekel, W.: *Multiple-flux-quanta vortices and quasiparticle states in the two-band superconductor Pb* arXiv:2211.08271 [cond-mat]. (2022). doi:10.48550/arXiv.2211.08271 (cit. on p. 97).

[208] Saunderson, T. G., Annett, J. F., Újfalussy, B., Csire, G. & Gradhand, M.: *Gap anisotropy in multiband superconductors based on multiple scattering theory*. Phys. Rev. B **101**, 064510. doi:10.1103/PhysRevB.101.064510 (2020) (cit. on p. 98).

[209] Choy, T.-S., Naset, J., Hershfield, S., Stanton, C. & Chen, J.: *A Database of Fermi Surfaces in Virtual Reality Modeling Language*, L36.042 (2000) (cit. on p. 98).

[210] Chanin, G. & Torre, J. P.: *Critical-Field Curve of Superconducting Lead*. Phys. Rev. B **5**, 4357–4364. doi:10.1103/PhysRevB.5.4357 (1972) (cit. on p. 98).

[211] Prozorov, R.: *Topological hysteresis in the intermediate state of type-I superconductors*. Phys. Rev. B **72**. doi:10.1103/PhysRevB.72.212508 (2005) (cit. on p. 98).

[212] Prozorov, R.: *Equilibrium Topology of the Intermediate State in Type-I Superconductors of Different Shapes*. Phys. Rev. Lett. **98**, 257001. doi:10.1103/PhysRevLett.98.257001 (2007) (cit. on pp. 98, 100).

[213] Prozorov, R., Fidler, A. F., Hoberg, J. R. & Canfield, P. C.: *Suprafroth in type-I superconductors*. Nature Physics **4**, 327–332. doi:10.1038/nphys888 (2008) (cit. on p. 98).

[214] Prozorov, R. & Hoberg, J. R.: *Dynamic formation of metastable intermediate state patterns in type-I superconductors*. J. Phys.: Conf. Ser. **150**, 052217. doi:10.1088/1742-6596/150/5/052217 (2009) (cit. on p. 98).

[215] Landau, L. D.: *The intermediate state of supraconductors*. Nat. **141**, 688 (1938) (cit. on p. 100).

[216] Landau, L. D.: *On the theory of the intermediate state of superconductors*. J. Phys. USSR **7**, 99 (1943) (cit. on p. 100).

[217] Hess, H. F., Robinson, R. B. & Waszczak, J. V.: *Vortex-core structure observed with a scanning tunneling microscope*. Phys. Rev. Lett. **64**, 2711–2714. doi:10.1103/PhysRevLett.64.2711 (1990) (cit. on p. 103).

[218] Hayashi, N., Ichioka, M. & Machida, K.: *Star-Shaped Local Density of States around Vortices in a Type-II Superconductor*. Phys. Rev. Lett. **77**, 4074–4077. doi:10.1103/PhysRevLett.77.4074 (1996) (cit. on p. 103).

[219] Hayashi, N., Ichioka, M. & Machida, K.: *Effects of gap anisotropy upon the electronic structure around a superconducting vortex*. Phys. Rev. B **56**, 9052–9063. doi:10.1103/PhysRevB.56.9052 (1997) (cit. on p. 103).

[220] Kim, H., Nagai, Y., Rózsa, L., Schreyer, D. & Wiesendanger, R.: *Anisotropic non-split zero-energy vortex bound states in a conventional superconductor*. Applied Physics Reviews **8**, 031417. doi:10.1063/5.0055839 (2021) (cit. on p. 103).

[221] Wong, W., Massa, M. & Nicol, E.: *Effect of Order Parameter Symmetry on the Local Density of States around an Isolated Vortex*. Journal of Low Temperature Physics **117**, 265–269. doi:10.1023/A:1022536709571 (1999) (cit. on p. 103).

[222] Kaneko, S.-i., Matsuba, K., Hafiz, M., Yamasaki, K., Kakizaki, E., Nishida, N., Takeya, H., Hirata, K., Kawakami, T., Mizushima, T. & Machida, K.: *Quantum Limiting Behaviors of a Vortex Core in an Anisotropic Gap Superconductor*. J. Phys. Soc. Jpn. **81**, 063701. doi:10.1143/JPSJ.81.063701 (2012) (cit. on p. 103).

[223] Gygi, F. & Schlüter, M.: *Self-consistent electronic structure of a vortex line in a type-II superconductor*. Phys. Rev. B **43**, 7609–7621. doi:10.1103/PhysRevB.43.7609 (1991) (cit. on pp. 103, 104, 108, 109).

[224] Kramer, L. & Pesch, W.: *Core structure and low-energy spectrum of isolated vortex lines in clean superconductors at $T \ll Tc$*. Z. Physik **269**, 59–64. doi:10.1007/BF01668869 (1974) (cit. on p. 104).

[225] Kramer, L., Pesch, W. & Watts-Tobin, R. J.: *Local structure and thermodynamic behavior of dirty superconductors in the mixed state at arbitrary temperature*. J Low Temp Phys **14**, 29–51. doi:10.1007/BF00654809 (1974) (cit. on p. 104).

[226] Doettinger, S. G., Huebener, R. P. & Kittelberger, S.: *Kramer-Pesch effect and damping of the vortex motion in the cuprate superconductors*. Phys. Rev. B **55**, 6044–6050. doi:10.1103/PhysRevB.55.6044 (1997) (cit. on p. 104).

[227] Volovik, G. E.: *Vortex motion in fermi-superfluids and Callan-Harvey effect.* Pis'ma v Zhurnal Ehksperimental'noj i Teoreticheskoj Fiziki; (Russian Federation) **57:3** (1993) (cit. on pp. 104, 110).

[228] Virtanen, S. M. M. & Salomaa, M. M.: *Multiquantum vortices in superconductors: Electronic and scanning tunneling microscopy spectra.* Phys. Rev. B **60**, 14581–14584. doi:10.1103/PhysRevB.60.14581 (1999) (cit. on pp. 108, 109).

[229] Caroli, C., De Gennes, P. G. & Matricon, J.: *Bound Fermion states on a vortex line in a type II superconductor.* Physics Letters **9**, 307–309. doi:10.1016/0031-9163(64)90375-0 (1964) (cit. on p. 108).

[230] Chen, M., Chen, X., Yang, H., Du, Z., Zhu, X., Wang, E. & Wen, H.-H.: *Discrete energy levels of Caroli-de Gennes-Matricon states in quantum limit in* $FeTe_{0.55}Se_{0.45}$. Nature Commun. **9**, 970. doi:10.1038/s41467-018-03404-8 (2018) (cit. on p. 109).

[231] Volovik, G. E.: *Localized fermions on quantized vortices in superfluid* ^{3}*He-B.* J. Phys.: Condens. Matter **3**, 357–368. doi:10.1088/0953-8984/3/3/010 (1991) (cit. on p. 110).

[232] Volovik, G. E.: *Topological Superfluids.* J. Exp. Theor. Phys. **129**, 618–641. doi:10.1134/S106377611910011X (2019) (cit. on p. 110).

[233] Bogomol'nyi, E. B.: *The stability of classical solutions.* Sov. J. Nucl. Phys. (Engl. Transl.); (United States) **24:4** (1976) (cit. on p. 113).

[234] Jacobs, L. & Rebbi, C.: *Interaction energy of superconducting vortices.* Phys. Rev. B **19**, 4486–4494. doi:10.1103/PhysRevB.19.4486 (1979) (cit. on p. 113).

[235] Weinberg, E. J.: *Multivortex solutions of the Ginzburg-Landau equations.* Phys. Rev. D **19**, 3008–3012. doi:10.1103/PhysRevD.19.3008 (1979) (cit. on p. 113).

[236] Taubes, C. H.: *ArbitraryN-vortex solutions to the first order Ginzburg-Landau equations.* Commun.Math. Phys. **72**, 277–292. doi:10.1007/BF01197552 (1980) (cit. on p. 113).

[237] Blatter, G., Feigel'man, M. V., Geshkenbein, V. B., Larkin, A. I. & Vinokur, V. M.: *Vortices in high-temperature superconductors.* Rev. Mod. Phys. **66**, 1125–1388. doi:10.1103/RevModPhys.66.1125 (1994) (cit. on pp. 114, 115).

[238] Schnyder, A. P., Ryu, S., Furusaki, A. & Ludwig, A. W. W.: *Classification of topological insulators and superconductors in three spatial dimensions.* Physical Review B **78**. doi:10.1103/PhysRevB.78.195125 (2008) (cit. on p. 117).

[239] Teo, J. C. Y. & Kane, C. L.: *Topological defects and gapless modes in insulators and superconductors.* Phys. Rev. B **82**, 115120. doi:10.1103/PhysRevB.82.115120 (2010) (cit. on p. 117).

[240] Hull, D. & Bacon, D. J.: *Introduction to dislocations* (Butterworth-Heinemann, 2011) (cit. on p. 119).

[241] Kadoyoshi, T., Kaburaki, H., Shimizu, F., Kimizuka, H., Jitsukawa, S. & Li, J.: *Molecular dynamics study on the formation of stacking fault tetrahedra and unfaulting of Frank loops in fcc metals.* Acta Materialia **55**, 3073–3080. doi:10.1016/j.actamat.2007.01.010 (2007) (cit. on pp. 119, 120).

[242] Schäublin, R., Yao, Z., Baluc, N. & Victoria, M.: *Irradiation-induced stacking fault tetrahedra in fcc metals.* Philosophical Magazine **85**, 769–777. doi:10.1080/14786430412331319929 (2005) (cit. on p. 119).

[243] Schouteden, K. & Van Haesendonck, C.: *Lateral Quantization of Two-Dimensional Electron States by Embedded Ag Nanocrystals.* Phys. Rev. Lett. **108**, 076806. doi:10.1103/PhysRevLett.108.076806 (2012) (cit. on pp. 119, 122, 125).

[244] Schouteden, K., Amin-Ahmadi, B., Li, Z., Muzychenko, D., Schryvers, D. & Van Haesendonck, C.: *Electronically decoupled stacking fault tetrahedra embedded in Au(111) films.* Nat Commun **7**, 14001. doi:10.1038/ncomms14001 (2016) (cit. on pp. 119, 120, 122).

[245] Müller, M., Néel, N., Crampin, S. & Kröger, J.: *Lateral Electron Confinement with Open Boundaries: Quantum Well States above Nanocavities at Pb(111).* Phys. Rev. Lett. **117**, 136803. doi:10.1103/PhysRevLett.117.136803 (2016) (cit. on pp. 121, 122, 125).

[246] Müller, M., Néel, N., Crampin, S. & Kröger, J.: *Open-boundary reflection of quantum well states at Pb(111).* Phys. Rev. B **96**, 205426. doi:10.1103/PhysRevB.96.205426 (2017) (cit. on pp. 122, 125).

[247] Song, S. Y. & Seo, J.: *Observation of enhanced superconductivity in the vicinity of Ar-induced nano-cavities in Pb(111).* Scientific Reports **7**, 12177. doi:10.1038/s41598-017-12505-1 (2017) (cit. on p. 124).

[248] McElroy, K., Simmonds, R. W., Hoffman, J. E., Lee, D.-H., Orenstein, J., Eisaki, H., Uchida, S. & Davis, J. C.: *Relating atomic-scale electronic phenomena to wave-like quasiparticle states in superconducting $Bi_2Sr_2CaCu_2O_{8+\delta}$.* Nature **422**, 592–596. doi:10.1038/nature01496 (2003) (cit. on p. 125).

[249] Vishik, I. M., Nowadnick, E. A., Lee, W. S., Shen, Z. X., Moritz, B., Devereaux, T. P., Tanaka, K., Sasagawa, T. & Fujii, T.: *A momentum-dependent perspective on quasiparticle interference in $Bi_2Sr_2CaCu_2O_{8+\delta}$.* Nature Phys **5**, 718–721. doi:10.1038/nphys1375 (2009) (cit. on p. 125).

[250] Li, W.-K. & Blinder, S. M.: *Solution of the Schrödinger equation for a particle in an equilateral triangle.* J. Math. Phys. **26**, 2784–2786. doi:10.1063/1.526701 (1985) (cit. on pp. 125, 126).

[251] Krishnamurthy, H. R., Mani, H. S. & Verma, H. C.: *Exact solution of the Schrodinger equation for a particle in a tetrahedral box*. J. Phys. A: Math. Gen. **15**, 2131–2137. doi:10.1088/0305-4470/15/7/024 (1982) (cit. on p. 125).

[252] Civale, L.: *Vortex pinning and creep in high-temperature superconductors with columnar defects*. Supercond. Sci. Technol. **10**, A11–A28. doi:10.1088/0953-2048/10/7A/003 (1997) (cit. on p. 127).

[253] Pi, U. H., Khim, Z. G., Kim, D. H., Schwarz, A., Liebmann, M. & Wiesendanger, R.: *Direct observation of vortices trapped at stacking fault dislocations in $Bi_2Sr_2CaCu_2O_8$ by a low-temperature magnetic force microscope*. Phys. Rev. B **69**, 094518. doi:10.1103/PhysRevB.69.094518 (2004) (cit. on p. 127).

[254] Morháč, M. & Matoušek, V.: *Complete positive deconvolution of spectrometric data*. Digital Signal Processing **19**, 372–392. doi:10.1016/j.dsp.2008.06.002 (2009) (cit. on p. 175).

[255] Eilenberger, G.: *Transformation of Gorkov's equation for type II superconductors into transport-like equations*. Z. Physik **214**, 195–213. doi:10.1007/BF01379803 (1968) (cit. on p. 179).

[256] Larkin, A. I. & Ovchinnikov, Y. N.: *Quasiclassical Method in the Theory of Superconductivity*. Sov. Phys. JETP **28**, 1200 (1969) (cit. on p. 179).

[257] Schopohl, N.: *Transformation of the Eilenberger Equations of Superconductivity to a Scalar Riccati Equation* arXiv:cond-mat/9804064. (1998). doi:10.48550/arXiv.cond-mat/9804064 (cit. on pp. 179, 181).

[258] Carneiro, G. & Brandt, E. H.: *Vortex lines in films: Fields and interactions*. Phys. Rev. B **61**, 6370–6376. doi:10.1103/PhysRevB.61.6370 (2000) (cit. on p. 180).

[259] Schopohl, N. & Maki, K.: *Quasiparticle spectrum around a vortex line in a d-wave superconductor*. Phys. Rev. B **52**, 490–493. doi:10.1103/PhysRevB.52.490 (1995) (cit. on pp. 180, 181).

Acronyms

AC Alternating current

ADC Analog-to-digital converter

AES Auger electron spectroscopy

AFV Antiferromagnetic ordering vector

ARPES Angle-resolved photoemission spectroscopy

BCS Bardeen-Cooper-Schrieffer

BdG Bogoliubov-de Gennes

BEC Bose-Einstein-Condensate

Bi2212 $Bi_2Sr_2CaCu_2O_{8+\delta}$

BSCCO Bismuth strontium calcium copper oxide

BZ Brillouin zone

CdGM Caroli-de Gennes-Matricon

D-STM Dilution scanning tunnelling microscope

DAC Digital-to-analog converter

DC Direct current

DFT Density functional theory

DOS Density of states

FFT Fast Fourier transform

FIM Field ion microscopy

FPGA Field programmable gate array

FS Fermi surface

FWHM Full width at half maximum

GL Ginzburg-Landau

grAl Granular aluminium

HTS High temperature superconductor

IETS Inelastic tunnelling spectroscopy

INS Inelastic neutron scattering

IQMT Institute for Quantum Materials and Technologies

irrep Irreducible representation

JJ Josephson junction

JT-STM Joule-Thomson scanning tunnelling microscope

LBCO Lanthanum barium copper oxide

LDOS Local density of states

LEED Low energy electron diffraction

MBE Molecular beam epitaxy

MIT Metal-to-insulator transition

MO Magneto-optical

MZM Majorana zero mode

NEG Non-evaporable getter

NG Nambu-Goldstone

NIN Normal metal-insulator-normal metal

NIS Normal metal-insulator-superconductor

NMR Nuclear magnetic resonance

QMS Quadrupole mass spectrometer

QP Quasiparticle

QPI Quasiparticle interference

QWS Quantum well state

RC Real-time controller

REBCO Rare earth barium copper oxide

RMS Root mean square

SCBA Self-consistent Born approximation

SEM Scanning electron microscopy

SFT Stacking fault tetrahedron

SIS Superconductor-insulator-superconductor

SIT Superconductor-to-insulator transition

SQUID Superconducting quantum interference device

STM Scanning tunnelling microscope

STS Scanning tunnelling spectroscopy

TMP Turbomolecular pump

TSP Titanium sublimation pump

TT Thompson tetrahedron

UHV Ultra-high vacuum

WKB Wentzel–Kramers–Brillouin

Y123 $YBa_2Cu_3O_{6+x}$

YBCO Yttrium barium copper oxide

YSR Yu-Shiba-Rusinov

ZFC Zero field cooling

Notation

Throughout this thesis, the SI unit system and the following notations were used:

- $\hbar = k_B = 1$ unless stated otherwise.
- $\beta = 1/(k_B T)$ denotes the inverse energy scale of the physical temperature.
- $n_{F/B}(x) = \frac{1}{\mathrm{e}^{-\beta x} \pm 1}$ denotes the Fermi-Dirac/Bose-Einstein distribution function.
- $\Theta(x) = \begin{cases} 1 & , x \geq 0 \\ 0 & , x < 0 \end{cases}$ denotes the Heaviside function.
- $A \otimes B$ denotes a Kronecker product of two vectors or matrices.

List of physical constants

- e, electron charge
- k_B, Boltzmann constant
- $\hbar$, reduced Planck constant
- c, vacuum speed of light
- m_e, electron rest mass
- ϵ_0, vacuum permittivity
- μ_0, vacuum permeability
- Φ_0, magnetic flux quantum

List of Figures

2.1 Ground State in GL Theory 6
2.2 Intermediate and Mixed Phase 8
2.3 Anderson-Higgs Mechanism 10
2.4 Electron-Electron Interaction 16
2.5 Mott MIT 21
2.6 Pairing Symmetries 25
2.7 (Para)magnon Exchange 29
2.8 Spin-Fermion Model's Spin Resonance 30
2.9 Tunnelling in STM and Planar Junctions 31
2.10 Elastic Tunnelling Processes 34
2.11 Signatures of Inelastic Tunnelling 36

3.1 Constant Current Mode 39
3.2 Energy Resolution Limit 42
3.3 D-STM Cryostat and STM 45
3.4 Tip and Sample Preparation 49

4.1 Structure Models of Bi2212 and Y123 52
4.2 Electronic Properties of Cuprates 53
4.3 Spin Excitation in INS on Bi2212 and Y123 54
4.4 Elastic and Inelastic Contributions 57
4.5 Conductance Inhomogeneity on Superconducting Bi2212 and Y123 . 59
4.6 Symmetrized Spectra of Superconducting Bi2212 and Y123 60
4.7 Normal State Electrons in Bi2212 and Y123 63
4.8 Spin Spectrum Extraction 64
4.9 Numerical Scaling Factor α 66

5.1 Superconducting Dome 70
5.2 Sample Structure 72
5.3 Pure Al Sample 74
5.4 Oxygen Poor Sample 75
5.5 Josephson Tunnelling 76
5.6 Oxygen Rich Sample 77

5.7 Grain Coupling 78
5.8 YSR States 79
5.9 Inelastic Excitations 81
5.10 Boson Spectral Density 84
5.11 Temperature Dependence of Low Energy Spectrum 86
5.12 LC Resonator Model 88
5.13 Spontaneous Plasmon Emission 91
5.14 Oxidized Al Film 93

6.1 Superconducting State 98
6.2 Intermediate State 99
6.3 Single Flux Vortex 102
6.4 Coherence Lengths in the Normal Single-flux Vortex 104
6.5 Anomalous Single-flux Vortex 106
6.6 Manipulation of CdGM State Displacements 107
6.7 Manipulation from Anomalous to Normal Vortex 107
6.8 CdGM states 108
6.9 Multiquantum Vortex Fingerprint 109
6.10 Multiflux Vortices 111
6.11 Giant Vortex 112
6.12 Low Temperature Transitional Region 113
6.13 Vortex at $T \lesssim T_c$ 114
6.14 Vortex Pinning Mechanisms 115

7.1 Flux Traps 117
7.2 Thompson Tetrahedron 118
7.3 Silcox-Hirsch Mechanism 119
7.4 Appearance of SFTs 121
7.5 Band Filter Effect 123
7.6 Scattering Vectors 124
7.7 Free Particle in an Equilateral Triangle Box 126
7.8 Single-flux Vortex in SFT 127
7.9 Edge State 128
7.10 Localisation and Lifetimes 129
7.11 Edge State in Buried and Surfacing SFTs 130
7.12 $m = 2$ Vortex in SFT 131

B.1 Inelastic Excitations 178
B.2 Boson Spectral Density 178

C.1 Riccati Parametrisation 182

C.2 Influence of Vector Potential 183

D.1 Reversal of Current/Magnetic Field Direction 185
D.2 Kramer-Pesch Effect 186

E.1 SFT in Normal Phase 187
E.2 Missing Edge State 188

List of Tables

2.1 Spin-Singlet Pairing States with Even Parity in Tetragonal and Orthorhombic Crystals 25

4.1 Fit Parameters for Y123 and Bi2212 Superconducting Spectra 63

5.1 (Granular) Aluminium Samples 73
5.2 Fit Parameters for Bosonic Excitations 82

B.1 Fit Parameters for Bosonic Excitations 178

Appendices

A Deconvolution using the Gold Algorithm

On a very general basis, the spectroscopic output signal of a system y with a certain response function h can be written as the discrete convolution

$$y(\omega_i) = \sum_j^i x(\omega_j)h(\omega_i - \omega_j) + n(\omega_i), \tag{A.1}$$

where x is the input signal, n is arbitrary noise and ω_i are discrete increments of frequency. In Chap. 4, $y \leftrightarrow \sigma^{\mathrm{i}}$, $x \leftrightarrow g^2\chi''$ and $h \leftrightarrow \sigma^{\mathrm{el}}$. One can write Eq. (A.1) in matrix form [254]:

$$\boldsymbol{y} = H\boldsymbol{x} + \boldsymbol{n}. \tag{A.2}$$

A least square solution of these coupled linear equations is found by minimization of the functional [254]

$$\left\| H\hat{\boldsymbol{x}} - \boldsymbol{y} \right\|_2 , \tag{A.3}$$

where $\hat{\boldsymbol{x}} = A^{-1}H\boldsymbol{y}$ is the unconstrained least squares estimate of the input signal and $A = H^TH$ is called the Toeplitz matrix. This solution is not definite, because the problem of finding x with knowledge of H and y is an ill-posed problem. There exist many solutions to this system of equations and small errors in assumption of H or noise cause large oscillations in the extracted solution. Regularization techniques are necessary to confine the possible solutions to a smaller subset. Common regularization techniques concentrate on modelling the noise. In the problem from Chap. 4, the shape of the noise is unknown but even worse, the input signal x is only a vague guess. Therefore, larger errors due to the uncertainty of H are expected to influence the deconvolution to find x. A physical constraint that drastically reduces the amount of possible solutions is the required positiveness of y, x and therefore also the positive definiteness of the Toeplitz matrix A.

The Gold algorithm [141, 254] is one of the few deconvolution algorithms that makes use of this constraint and always yields positive solutions if the input is positive. In the classic Gold algorithm, the solution is refined iteratively by the procedure [254]

$$x^{(n+1)}(\omega_i) = \frac{y'(\omega_i)}{\sum_{j=0}^{M-1} A_{ij}x^{(n)}(\omega_j)} x^{(n)}(\omega_i). \tag{A.4}$$

This solution converges to the least square estimate in the constrained subspace of positive solutions and is frequently used in the analysis of γ-spectroscopy data where detector counts are necessarily positive. The algorithm also decomposes the solution into a series of δ-peaks which is not ideal for the problem at hand, but can pronounce the seeked bosonic mode. Since the dimension N of A takes at least the number of points in the spectrum, it is large and the product of Toeplitz matrix and vector $\boldsymbol{x}$ in Eq. (A.4) is done by the FFT. In fact, the kernel function is far from periodic and does not naturally go to zero for $\omega \to \infty$ which is why it has to reasonably padded, meaning it is first extrapolated and then smoothly brought to zero. The total deconvolution range then spans a range much larger than the range where the kernel is unequal to zero. A typical total size of the Toeplitz matrix is $N = 12000$ due to the extensive padding that is required. The FFT thus reduces the computation time for one step by a factor of

$$\frac{N^2}{\frac{1}{2} N \log_2 N} \approx 1770. \tag{A.5}$$

B Supplementary GrAl Data

In this appendix, supplementary data supporting the conclusion from Sec. 5.4 is presented. Fig. B.1 shows a spectrum containing out-gap peaks indicating inelastic excitations of the tunnelling electron, just like in Fig. 5.9 of the main text. This spectrum was obtained at a different tip position and slightly milder tunnelling conditions, so smaller overall conductance at the feedback condition ($U = 4\,\mathrm{mV}$, $I = 70\,\mathrm{pA}$). It was thoroughly checked that periodic noise in the tunnelling contact is not causing the equidistant peaks in the spectrum. For good measure, the bias step size was changed (while keeping the integration time for each bias value equal) compared to the spectrum of Fig. 5.9. The spectrum shown here is an average of 1000 single spectra. For the analysis of this spectrum, the model from Eq. (5.2) of the main text was fitted to the data. Tab. B.1 shows the found parameters, when fitting first the negative bias side for the energies and then imposing them for the positive side. Again, with five excitations, the spectrum is qualitatively well reproduced (see red line in Fig. B.1). Especially the ω_n for $n = 2, 3, 4$ are very similar to the ω_n found for the data presented in the main text. From this analysis, ω_1 is slightly larger than $1.35\,\Delta$.

The deconvolution by the FFT and Gold algorithm, however, reveals peaks in the boson spectral density $\alpha^2 F(\omega)$ at multiples of $\omega_{\mathrm{exc}} = 1.35\,\Delta$, as indicated by the grey dashed lines in the bottom panel of Fig. B.2. Solely the fourth excitation is found at higher energies than expected. The trend of increasing amplitudes for harmonics of $n < 3$ and subsequent decrease for harmonics of $n > 3$ is in good agreement with what was interpreted into the boson spectral density obtained in the main text.

Table B.1: Fit Parameters for Bosonic Excitations: Fitted parameters of Eq. (5.2) used to model the spectrum in red in Fig. B.1. n is the number of the bosonic excitation, ω_n its energy divided by $\Delta = (250 \pm 3)$ µeV and a_n^+ and a_n^- are the intensities for positive and negative bias range.

n	ω_n (Δ)	a_n^+	a_n^-
1	1.52	0.247	0.533
2	2.75	1.352	0.258
3	4.16	1.114	0.892
4	5.18	0.000	1.053
5	6.08	0.975	0.803

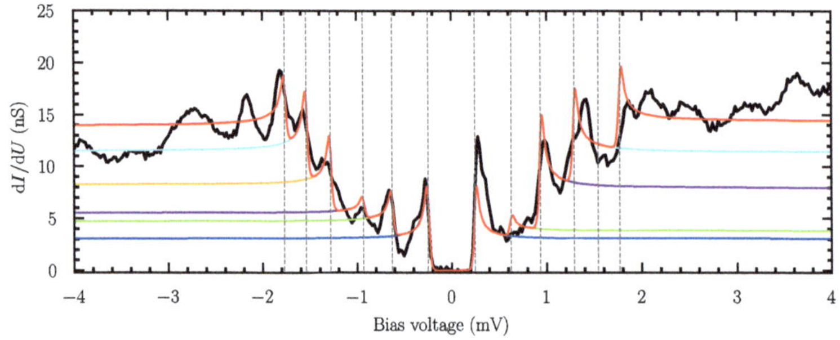

Figure B.1: Inelastic Excitations: Analogon to Fig. 5.9, recorded at a different tip position. Here, the left side was used for the determination of the five excitation energies. Similar energies as in the main text are found (see Tab. B.1).

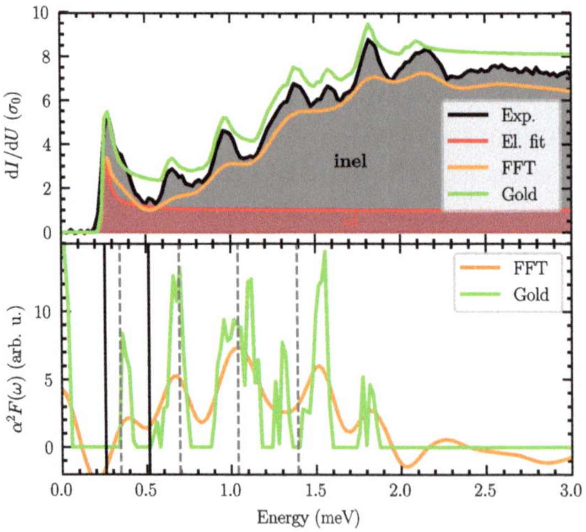

Figure B.2: Boson Spectral Density: Analogon to Fig. 5.10 using the spectrum in Fig. B.1 as a source. Grey dashed lines indicate the energies $n \cdot \omega_{\text{exc}}$ for $n = 1, 2, 3, 4$ and $\omega_{\text{exc}} = 1.35\,\Delta$ like in the main text.

C 3D Eilenbenberger Equations

Gert Eilenberger [255] and by a slightly different approach Larkin and Ovchinnikov [256] could show that the Gorkov equations of motion for normal and anomalous Green's functions can be transformed into a set of simpler transport-like equations for quasi-classical Green's function propagators, that hold for $k_F\xi \gg 1$ [257]:

$$-\hbar \boldsymbol{v}_F \nabla \hat{g}(\boldsymbol{r};\boldsymbol{p}_F,i\epsilon_n) = \left[\begin{pmatrix} i\epsilon_n + \boldsymbol{v}_F e\boldsymbol{A}(\boldsymbol{r}) & -\Delta(\boldsymbol{r},\boldsymbol{p}_F) \\ \Delta^\dagger(\boldsymbol{r},\boldsymbol{p}_F) & -i\epsilon_n - \boldsymbol{v}_F e\boldsymbol{A}(\boldsymbol{r}) \end{pmatrix}, \hat{g}(\boldsymbol{r};\boldsymbol{p}_F,i\epsilon_n)\right]. \tag{C.1}$$

The quasi-classical Green's function propagator

$$\hat{g} = \begin{pmatrix} g(\boldsymbol{r},\boldsymbol{p}_F,i\epsilon_n) & f(\boldsymbol{r},\boldsymbol{p}_F,i\epsilon_n) \\ -f^*(\boldsymbol{r},-\boldsymbol{p}_F,i\epsilon_n) & g^*(\boldsymbol{r},-\boldsymbol{p}_F,i\epsilon_n) \end{pmatrix}, \tag{C.2}$$

depends on spatial coordinate $\boldsymbol{r}$, crystal momentum $\boldsymbol{p}_F = \hbar\boldsymbol{k}_F$ and energy ϵ_n, and must satisfy the normalization condition $\hat{g}^2 = \hat{1}$. g and f are normal and anomalous quasi-classical Green's function propagators and ϵ_n are fermionic Matsubara frequencies. Incorporated in this formalism is the self-consistent calculation of pair potential $\Delta(\boldsymbol{r})$ and $\boldsymbol{A}(\boldsymbol{r})$, which is essential for the description of a superconductor hosting vortices as these two fields are dependent on each other. Although this approach is numerically much more feasible than direct diagonalization of a BdG Hamiltonian when treating inhomegeneties, it is not necessary to solve the problem completely self-consistently in this work. The reason for that is, that the experimental results from Chap. 6 clearly show, that, even though the sub-gap states in the vortex display highly anisotropic behaviour, the recovery of $\Delta(\boldsymbol{r})$ is isotropic in the plane around the vortex core. The local pair potential is assumed to have s-wave symmetry and has therefore been modeled by

$$\Delta(\boldsymbol{r},\boldsymbol{p}_F) = \Delta(\boldsymbol{r}) = \left(\Delta_0 \tanh\frac{r}{\xi} + \Theta(z) W \tanh\frac{z}{a}\right)\left(\frac{x+iy}{r}\right)^m, \tag{C.3}$$

with Θ being a Heaviside step function, Δ_0 the maximum gap size, ξ the coherence length, a the lattice constant, m the winding number of the vortex and W the work function. The second term ensures, that quasiparticles travelling to the surface are decaying into the vacuum with the right damping factor. Since an isotropic Δ implies an isotropic in-plane current density, the magnetic field profile around the vortex is described by a vector potential of the form [258]

$$\boldsymbol{A} = A(r,z)\hat{\boldsymbol{e}}_\varphi\,, \tag{C.4}$$

$$A(r,z) = \frac{m\Phi_0}{2\pi\lambda^2}\int\limits_0^\infty \mathrm{d}k \frac{J_1(kr)}{k^2+\lambda^{-2}} S(k,z)\,, \tag{C.5}$$

$$S(k,z) = \begin{cases} \frac{\kappa}{k+\kappa}\mathrm{e}^{-kz} & z>0 \\ 1-\frac{k}{k+\kappa}\mathrm{e}^{\kappa z} & z\leq 0\,, \end{cases} \tag{C.6}$$

that is cylinder symmetric in the bulk and deviates from the bulk value near and above the surface. Here, $\kappa = \sqrt{k^2+\lambda^{-2}}$, λ is the magnetic penetration depth which was chosen to be $\lambda = \xi/\sqrt{2}$ and $J_1(x)$ is a Bessel function of first order.

Nils Schopohl and Kazumi Maki [259] could show that the Eilenberger equations can always be solved along a characteristic line and that the solution is universal for every point along this line. The line simply has to be parallel to the Fermi velocity vector $\boldsymbol{v}_F$. Points on this line are then characterized by the variable X and two *impact parameters* Y and Z.

$$\boldsymbol{r}(X) = X\hat{\boldsymbol{u}} + Y\hat{\boldsymbol{v}} + Z\hat{\boldsymbol{w}}, \tag{C.7}$$

$$= x\hat{\boldsymbol{x}} + y\hat{\boldsymbol{y}} + z\hat{\boldsymbol{z}} \tag{C.8}$$

Transformation from the mobile frame of reference $(\hat{\boldsymbol{u}}, \hat{\boldsymbol{v}}, \hat{\boldsymbol{w}})$, where $\hat{\boldsymbol{u}} \parallel \boldsymbol{v}_F$, to the fixed coordinate system $(\hat{\boldsymbol{x}}, \hat{\boldsymbol{y}}, \hat{\boldsymbol{z}})$ is done by chaining rotation matrices using Euler angles:

$$\begin{pmatrix} v_x \\ v_y \\ v_z \end{pmatrix} = R_{zyz} \begin{pmatrix} v_X \\ v_Y \\ v_Z \end{pmatrix}\,, \tag{C.9}$$

$$\begin{aligned} R_{zyz} &= R_z(\eta)R_y(\chi-\pi/2)R_z(\psi) \\ &= \begin{pmatrix} \cos\eta & -\sin\eta & 0 \\ \sin\eta & \cos\eta & 0 \\ 0 & 0 & 1 \end{pmatrix} \begin{pmatrix} \sin\chi & 0 & \cos\chi \\ 0 & 1 & 0 \\ -\cos\chi & 0 & \sin\chi \end{pmatrix} \begin{pmatrix} \cos\psi & -\sin\psi & 0 \\ \sin\psi & \cos\psi & 0 \\ 0 & 0 & 1 \end{pmatrix}. \end{aligned} \tag{C.10}$$

In order to align the axis $\hat{\boldsymbol{u}}$ of the mobile frame with the axis $\hat{\boldsymbol{x}}$ of the static coordinate system, no rotation about $\hat{\boldsymbol{w}}$ is needed and thus $\psi = 0$. The other two rotational angles are defined by the components of the Fermi velocity vector as follows:

$$\eta = \arccos \frac{v_z}{|\boldsymbol{v}|}, \tag{C.11}$$

$$\chi = \begin{cases} \arctan \frac{v_y}{v_x}, & v_x > 0 \\ \arctan \frac{v_y}{v_x} + \pi, & v_x < 0 \,. \\ \operatorname{sgn}(v_y) \frac{\pi}{2}, & v_x = 0 \end{cases} \tag{C.12}$$

Using this transformation, the vector potential and gap parameter can be expressed in the variables of the mobile frame: $A(\boldsymbol{r}(X, Y, Z))$ and $\Delta(\boldsymbol{r}(X, Y, Z))$. On a characteristic line defined by the impact parameters Y_p and Z_p, the functions from Eq.(C.1) become [257]

$$\Delta(X) = \Delta(\boldsymbol{r}(X, Y_p, Z_p)), \tag{C.13}$$

$$i\tilde{\epsilon}_n(X) = i\epsilon_n + \boldsymbol{v}_F e\boldsymbol{A}(\boldsymbol{r}(X, Y_p, Z_p)), \tag{C.14}$$

$$\hat{g}(X) = \hat{g}(\boldsymbol{r}(X, Y_p, Z_p), \boldsymbol{p}_F, i\epsilon_n). \tag{C.15}$$

The parametrisation of the Eilenberger equations on this 1D line is called the *Riccati parametrisation*. Eq. (C.1) formally reduces to the solution of two initial value problems, where the differential equations are scalar and of first order. The Eilenberger propagator is parametrised in terms of two scalar complex functions $a(X)$ and $b(X)$:

$$\hat{g}(X) = \frac{1}{1 + a(X)b(X)} \begin{pmatrix} 1 - a(X)b(X) & 2ia(X) \\ -2ib(X) & -1 + a(X)b(X) \end{pmatrix}. \tag{C.16}$$

The differential equations that need to be numerically solved for $a(X)$ and $b(X)$ are [259]

$$\hbar v_F a'(X) + [2\tilde{\epsilon}_n + \Delta^\dagger(X)a(X)]a(X) - \Delta(X) = 0, \tag{C.17}$$

$$\hbar v_F b'(X) - [2\tilde{\epsilon}_n + \Delta(X)b(X)]b(X) + \Delta^\dagger(X) = 0 \tag{C.18}$$

with boundary conditions

$$a(-\infty) = \frac{\Delta(-\infty)}{\epsilon_n + \sqrt{\epsilon_n^2 + |\Delta(-\infty)|^2}}, \tag{C.19}$$

$$b(+\infty) = \frac{\Delta^\dagger(+\infty)}{\epsilon_n + \sqrt{\epsilon_n^2 + |\Delta(+\infty)|^2}}. \tag{C.20}$$

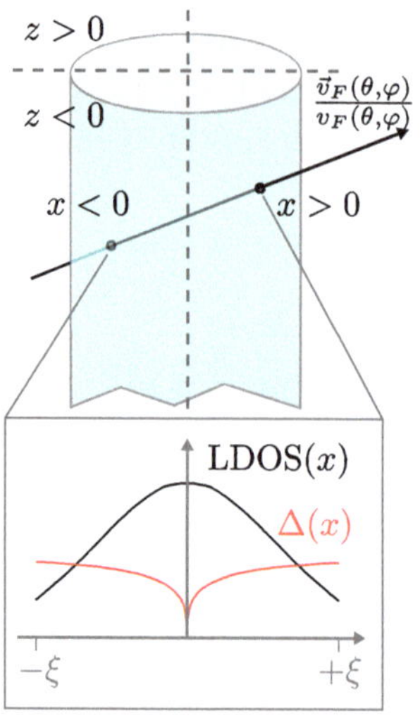

Figure C.1: Riccati Parametrisation: A sketch of the calculation procedure using the Riccati parametrization: The 3D trajectories (black arrow) through the flux tube (blue) are given by the Fermi velocities for each point $k_F(\theta, \varphi)$ on the Fermi surface. Solution of the effective 1D Eilenberger equations for this trajectory yields the LDOS(x) along this line. For a complete picture, a summation over all $k_F(\theta, \varphi)$ and impact parameters y is performed.

In order to solve them at a certain energy $E < \Delta_0$, the analytical continuation $i\epsilon_n \to E + i0^+$ is used. Finally, the local density of states at $\boldsymbol{r}$ is obtained from the real part of $g(\boldsymbol{r}(X))$, i.e.

$$\mathcal{N}(\boldsymbol{p}_F) = \mathcal{N}_0(\boldsymbol{p}_F)\mathrm{Re}\left(\frac{1 - a(X)b(X)}{1 + a(X)b(X)}\right). \tag{C.21}$$

In order to obtain the total local density of states, one still needs an integration over the Fermi surface to calculate all trajectories that are present due to the various Fermi velocity vectors that exist, plus an integration over the impact factors Y in order to account for trajectories that do not traverse the vortex centre. An integration over Z is redundant since the solution in the bulk is the same for every Z. Near the surface, this is strictly not the case anymore but the effect is small and with a large enough variety of Fermi velocity vectors (which we have) these trajectories are not missed. The above described calculation procedure in the Riccati parametrisation is schematically sketched in Fig. C.1.

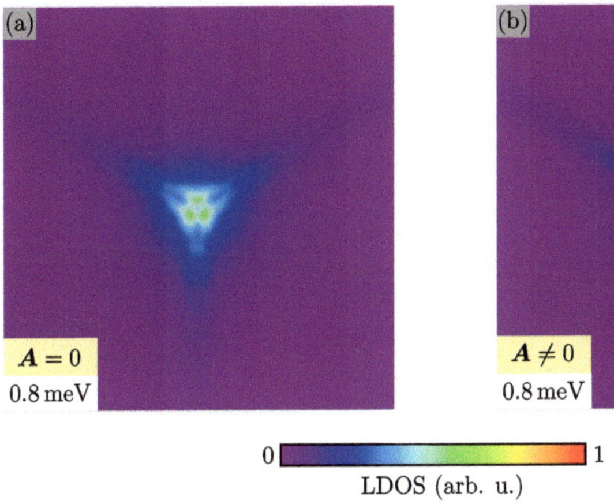

Figure C.2: Influence of Vector Potential: The LDOS obtained from solutions of the 3D Eilenberger equations for a single-flux vortex at energy $eU = 0.8\,\mathrm{meV}$ without (a) and with vector potential (b), as formulated in Eq. (C.6), shows only quantitative differences. With non-zero vector field, the star arms are still split, yet the CdGM states at this energy are squeezed into a smaller area around the vortex core.

C.1 Intermediate Results

The inclusion of a non-zero vector potential $\boldsymbol{A}$ in the calculations increases the average time needed to solve the Eilenberger equations because the Matsubara frequencies in Eq. (C.14) gain a position dependent term that recquires a coordinate transformation between the two reference frames. Therefore, the simulations shown in the main text are performed without vector potential. It is already visible from Eq. (C.6) that in a calculation where $\boldsymbol{A}$ and Δ are not solved self-consistently, $\boldsymbol{A}$ only enters the equation as an effective energy term. With a vector potential in the azimuthal direction, like in Eq. (C.6), the scalar product with the Fermi velocity is only expected to yield significant contribution for large impact parameters (for $Y_p = 0$, $\boldsymbol{v}_F$ is perpendicular to $\boldsymbol{A}$). That means trajectories with increasing impact parameter have large LDOS already for smaller distances from the core than in the field free case. The splitting star arms in the LDOS maps should be squeezed to smaller distances from the core. In fact, this is what is seen in the calculations with vector potential at higher energies, as shown in Fig. C.2. This proves that even though there are quantitative differences to the case without vector potential, in the general characteristics, the LDOS patterns remain unchanged.

D Supplementary Pb Vortex Data

D.1 Bias and Magnetic Field Dependence of CdGM LDOS

Under reversal of bias voltage sign, i.e. the tunnelling current direction ($I \leftrightarrow -I$), or magnetic field direction ($B\hat{e}_z \leftrightarrow -B\hat{e}_z$), the differential conductance maps show no qualitative change. As a point of proof, Fig. D.1(a,b) shows a normal vortex stabilized at negative field after exposing the sample to $-85\,\mathrm{mT}$ for positive voltages and (c,d) an anomalous vortex at $18\,\mathrm{mT}$ for negative voltages. This illustrates the electron-hole symmetry of the CdGM states and justifies treating them as excitations of a BCS ground state, i.e. describing their time-evolution by a mean-field BdG Hamiltonian.

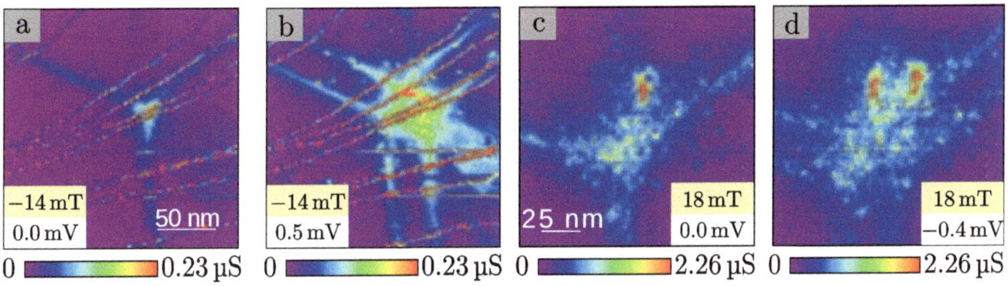

Figure D.1: Reversal of Current/Magnetic Field Direction: At a magnetic field in $-z$ direction (a,b) or at a negative bias voltage (c,d), the LDOS pattern of normal and anomalous vortices does not change qualitatively compared to the patterns shown in Chap. 6.

D.2 Self-consistent Solution of the Gap Equation

The self-consistent solution of the gap equation for a vortex with isotropic Fermi velocity in Fig. D.2 shows that a Kramer-Pesch effect is to be expected in this case: For a radius from the vortex centre of $r < 2\xi_0$ the recovery of $\Delta(r)$ significantly deviates from the function $\Delta_0 \tanh(r/\xi_0)$. The slope of $\Delta(r)$ near the centre is roughly twice as steep as expected from a tanh behaviour with universal ξ_0. This leads to a core size $\xi^{(c)} = \left[\lim_{r\to 0} \frac{\mathrm{d}\Delta(r)}{\mathrm{d}r}\right]^{-1}$ that is roughly half as big.

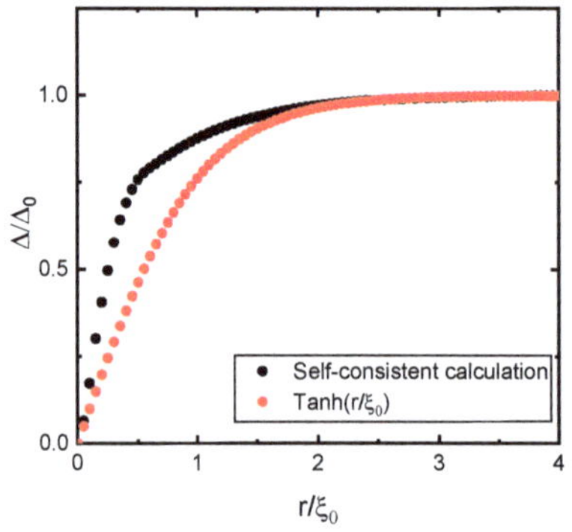

Figure D.2: Kramer-Pesch Effect: The self-consistent calculation of the pair potential Δ for a vortex with isotropic Fermi velocity exhibits a shrinking of its core size according to the Kramer-Pesch effect, i.e. a steeper recovery of Δ close to the vortex centre that does not follow $\tanh(r/\xi_0)$.

E Supplementary Pb Defect Data

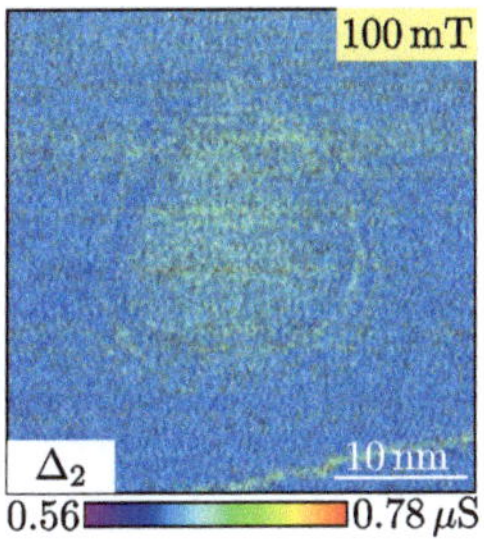

Figure E.1: SFT in Normal Phase: $\mathrm{d}I/\mathrm{d}U$ map ($eU = \Delta_2$) of the defect in Fig.7.4(a,e) at $B_z = 100\,\mathrm{mT}$ (normal conducting phase). No QPI is visible.

In the normal state at $B > B_c$, the SFT does not exhibit the characteristic LDOS pattern known from the superconducting state at $eU = \Delta_2$, as Fig. E.1 shows. A QPI pattern is also missing.

In the superconducting state, without pinned flux, the SFT misses the zero energy edge state. As Fig. E.2 shows, the superconducting gap is free of states in this region.

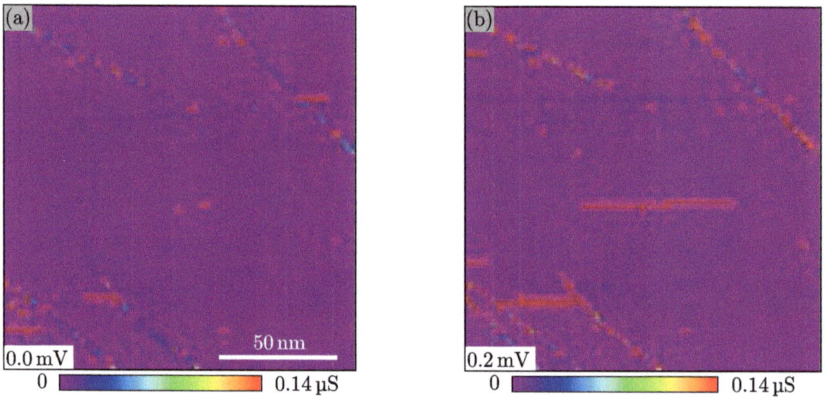

Figure E.2: Missing Edge State: $\mathrm{d}I/\mathrm{d}U$ maps at $U = 0.0\,\mathrm{mV}$ (a) and $U = 0.2\,\mathrm{mV}$ (b) on the SFT from Fig. 7.4(c) in the superconducting state at $B = 31\,\mathrm{mT}$ with no pinned vortex. An in-gap edge state of the SFT like in Fig. 7.8(a) and Fig. 7.9(a) is missing.

Acknowledgements

Of course, this thesis would not have been possible without the support, encouragement and guiding by many wonderful people. I am deeply grateful to my supervisor **Prof. Wulf Wulfhekel** for giving me the opportunity to do my doctorate on such an interesting topic. I highly appreciate his continuous support as well as the scientific discussions, the supervision and the shared ideas on such a frequent basis.

I would like to thank **Prof. Jörg Schmalian** for being the second referee, for many helpful discussions and the collaboration on joint publications.

Thanks also go to our staff for always having our backs, so we can concentrate on our research. Special thanks go to **Steffi Baatz**, **Claudia Alaya**, **Michael Meyer** and **Jannis Ret**.

I have to thank my colleague **Dr. Qili Li** for his tireless effort both in the lab and in computational matters. I also thank **Dr. Rolf Heid** for the DFT calculations and many fruitful discussions. Speaking of discussions, I am very grateful to **Dr. Roland Willa**, **Dr. Ioan Pop** and **Prof. Alexander Shnirman** for the discussions and their help in related fields of physics. Special thanks also go to **Dr. Fang Yang** for the hard work on our joint publication and to him and **Dr. Jasmin Jandke** for the great preliminary work.

I want to thank my students **Mirjam Henn** and **Janic Beck** for their interest in my field of study, the great teamwork and the fruitful discussions.

I am lucky to have had many excellent physicists as my teachers throughout the years. I thank you for all your valuable lessons in and out of the lab, **Dr. Timofey Balashov**, **Dr. Toshio Miyamachi**, **Dr. Timo Frauhammer**, **Dr. Marie Hervé** and **Dr. Hung-Hsiang Yang**.

I really enjoyed my stay in the group and want to say thank you to each and everyone that made this time unforgettable. Thank you, **Wulf**, **Timofey**, **Lukas**, **Kevin**, **Marie**, **Fang**, **Sergey**, **Qili**, **Hung-Hsiang**, **Hongyan**, **Loïc**, **Julian**, **Vibhuti**, **Simon**, **Philipp**, **Mirjam**, **Gabriel**, **Qing**, **Haoran**, **Paul** and **Timo**.

Lastly, my sincerest gratitude goes to my wonderful friends and family for their incredible support at each step of the way.

Experimental Condensed Matter Physics
(ISSN 2191-9925)

Band 1	Alexey Feofanov **Experiments on flux qubits with pi-shifters.** ISBN 978-3-86644-644-1
Band 2	Stefan Schmaus **Spintronics with individual metal-organic molecules.** ISBN 978-3-86644-649-6
Band 3	Marc Müller **Elektrischer Leitwert von magnetostriktiven Dy-Nanokontakten.** ISBN 978-3-86644-726-4
Band 4	Torben Peichl **Einfluss mechanischer Deformation auf atomare Tunnelsysteme – untersucht mit Josephson Phasen-Qubits.** ISBN 978-3-86644-837-7
Band 5	Dominik Stöffler **Herstellung dünner metallischer Brücken durch Elektromigration und Charakterisierung mit Rastersondentechniken.** ISBN 978-3-86644-843-8
Band 6	Tihomir Tomanic **Untersuchung des elektronischen Oberflächenzustands von Ag-Inseln auf supraleitendem Niob (110).** ISBN 978-3-86644-898-8
Band 7	Lukas Gerhard **Magnetoelectric coupling at metal surfaces.** ISBN 978-3-7315-0063-6
Band 8	Nicht erschienen.
Band 9	Jochen Zimmer **Cooper pair transport in arrays of Josephson junctions.** ISBN 978-3-7315-0130-5
Band 10	Oliver Berg **Elektrischer Transport durch Nanokontakte von Selten-Erd-Metallen.** ISBN 978-3-7315-0209-8
Band 11	Grigorij Jur'evic Grabovskij **Investigation of coherent microscopic defects inside the tunneling barrier of a Josephson junction.** ISBN 978-3-7315-0210-4

Die Bände sind unter www.ksp.kit.edu als PDF frei verfügbar oder als Druckausgabe bestellbar.

Band 12 Cornelius Thiele
STM Characterization of Phenylene-Ethynylene Oligomers on Au(111) and their Integration into Carbon Nanotube Nanogaps.
ISBN 978-3-7315-0235-7

Band 13 Michael Peter Schackert
Scanning Tunneling Spectroscopy on Electron-Boson Interactions in Superconductors.
ISBN 978-3-7315-0238-8

Band 14 Susanne Butz
One-Dimensional Tunable Josephson Metamaterials.
ISBN 978-3-7315-0271-5

Band 15 Philipp Jung
Nonlinear Effects in Superconducting Quantum Interference Meta-Atoms.
ISBN 978-3-7315-0294-4

Band 16 Sebastian Probst
Hybrid quantum system based on rare earth doped crystals.
ISBN 978-3-7315-0345-3

Band 17 Wolfram Kittler
Magnetische Anisotropie und Quantenphasenübergang in $CeTi_{1-x}V_xGe_3$.
ISBN 978-3-7315-0363-7

Band 18 Moritz Peter
Towards magnetic resonance in scanning tunneling microscopy using heterodyne detection.
ISBN 978-3-7315-0410-8

Band 19 Junji Tobias Märkl
Investigation of Magnetic Adatoms with Scanning Tunneling Techniques.
ISBN 978-3-7315-0435-1

Band 20 Jochen Braumüller
Quantum simulation experiments with superconducting circuits.
ISBN 978-3-7315-0780-2

Band 21 Jinjie Chen
Local Investigation of Single Magnetic Molecules with Scanning Tunneling Microscopy.
ISBN 978-3-7315-0819-9

Band 22 Arnold Seiler
Einfluss der Leitungselektronen auf die Dynamik atomarer Tunnelsysteme in ungeordneten Festkörpern: Relaxationsprozesse in metallischen Gläsern und ungeordneten dünnen Aluminiumoxid-Schichten.
ISBN 978-3-7315-0870-0

Band 23 Jasmin Maria Jandke
Elastic and Inelastic Scanning Tunneling Spectroscopy on Iron-Based Superconductors.
ISBN 978-3-7315-0747-5

Band 24 Kevin Edelmann
Electroluminescence from Plasmonic Excitations in a Scanning Tunnelling Microscope.
ISBN 978-3-7315-0923-3

Band 25 Lei Zhang
Sub-Kelvin scanning tunneling microscopy on magnetic molecules.
ISBN 978-3-86644-950-3

Band 26 Lukas Grünhaupt
Granular aluminium superinductors.
ISBN 978-3-7315-0956-1

Band 27 Alexander Bilmes
Resolving locations of defects in superconducting transmon qubits.
ISBN 978-3-7315-0967-7

Band 28 Marco Pfirrmann
Adding nonlinearity to an electromagnetic-magnonic quantum hybrid device.
ISBN 978-3-7315-1003-1

Band 29 Andre Schneider
Quantum Sensing Experiments with Superconducting Qubits.
ISBN 978-3-7315-1032-1

Band 30 Vibhuti Narayan Rai
Light Emission from Single Self-decoupled Molecules in a Scanning Tunnelling Microscope.
ISBN 978-3-7315-1201-1

Band 31 Julian Antoni Skolaut
Molecular Motor Based on Single Chiral Tripodal Molecules Studied with STM.
ISBN 978-3-7315-1247-9

Band 32 Thomas Gozlinski
Local Probing of a Superconductor's Quasiparticles and Bosonic Excitations with a Scanning Tunnelling Microscope.
ISBN 978-3-7315-1279-0

Die Bände sind unter www.ksp.kit.edu als PDF frei verfügbar oder als Druckausgabe bestellbar.